D1236368

Digital Video Quality

Digital Video Quality

Vision Models and Metrics

Stefan Winkler

Genista Corporation, Montreux, Switzerland

John Wiley & Sons, Ltd

Other Wiley Editorial Offices

John Wiley & Sons Inc., 111 River Street, Hoboken, NJ 07030, USA

Jossey-Bass, 989 Market Street, San Francisco, CA 94103-1741, USA

Wiley–VCH Verlag GmbH, Boschstr. 12, D-69469 Weinheim, Germany

John Wiley & Sons Australia Ltd, 33 Park Road, Milton, Queensland 4064, Australia

John Wiley & Sons (Asia) Pte Ltd, 2 Clementi Loop #02-01, Jin Xing Distripark, Singapore 129809

John Wiley & Sons Canada Ltd, 22 Worcester Road, Etobicoke, Ontario, Canada M9W 1L1

Wiley also publishes its books in a variety of electronic formats. Some content that appears in
print may not be available in electronic books.

Library of Congress Cataloging-in-Publication Data

Winkler, Stefan.
    Digital video quality : vision models and metrics / Stefan Winkler.
        p. cm.
    Includes bibliographical references and index.
    ISBN 0-470-02404-6
1. Digital video.   2. Image processing—Digital techniques.   3. Imaging
systems—Image quality.   I. Title.
    TK6680.5.W55 2005
    006.6′96–dc22   2004061588

British Library Cataloguing in Publication Data

A catalogue record for this book is available from the British Library

    ISBN 0-470-02404-6

Typeset in 10.5/13pt Times by Thomson Press (India) Limited, New Delhi
Printed and bound in Great Britain by Antony Rowe Ltd, Chippenham, Wiltshire
This book is printed on acid-free paper responsibly manufactured from sustainable forestry
in which at least two trees are planted for each one used for paper production.

Contents

About the Author

O, what may man within him hide,
Though angel on the outward side!

William Shakespeare

Stefan Winkler was born in Horn, Austria. He received the M.Sc. degree with highest honors in electrical engineering from the University of Technology in Vienna, Austria, in 1996, and the Ph.D. degree in electrical engineering from the École Polytechnique Fédérale de Lausanne (EPFL), Switzerland, in 2000 for work on vision modeling and video quality measurement. He also spent one year at the University of Illinois at Urbana-Champaign as a Fulbright student. He did internships at Siemens, ROLM, German Aerospace, Andersen Consulting, and Hewlett-Packard.

In January 2001 he co-founded Genimedia (now Genista), a company developing perceptual quality metrics for multimedia applications. In October 2002, he returned to EPFL as a post-doctoral fellow, and he also held an assistant professor position at the University of Lausanne for a semester. Currently he is Chief Scientist at Genista Corporation.

Dr Winkler has been an invited speaker at numerous technical conferences and seminars. He was organizer of a special session on video quality at VCIP 2003, technical program committee member for ICIP 2004 and WPMC 2004, and has been serving as a reviewer for several scientific journals. He is the author and co-author of over 30 publications on vision modeling and quality assessment.

Acknowledgements

I thank you most sincerely for your assistance;
whether or no my book may be wretched,
you have done your best to make it less wretched.

Charles Darwin

The basis for this book was my PhD dissertation, which I wrote at the Signal Processing Lab of the École Polytechnique Fédérale de Lausanne (EPFL) under the supervision of Professor Murat Kunt. I appreciated his guidance and the numerous discussions that we had. Christian van den Branden Lambrecht, whose work I built upon, was also very helpful in getting me started. I acknowledge the financial support of Hewlett-Packard for my PhD research.

I enjoyed working with my colleagues at the Signal Processing Lab. In particular, I would like to mention Martin Kutter, Marcus Nadenau and Pierre Vandergheynst, who helped me shape and realize many ideas. Yousri Abdeljaoued, David Alleysson, David McNally, Marcus Nadenau, Francesco Ziliani and my brother Martin read drafts of my dissertation chapters and provided many valuable comments and suggestions for improvement. Professor Jean-Bernard Martens from the Eindhoven University of Technology gave me a lot of feedback on my thesis. Furthermore, I thank all the people who participated in my subjective experiments for their time and patience.

Kambiz Homayounfar and Professor Touradj Ebrahimi created Genimedia and thus allowed me to carry on my research in this field and to put my ideas into products; they also encouraged me to work on this book. I am grateful to all my colleagues at Genimedia/Genista for the stimulating discussions we had and for creating such a pleasant working environment.

Thanks are due to the anonymous reviewers of the book for their helpful feedback. Simon Robins spent many hours with painstaking format conversions and more proofreading. I also thank my editor Simone Taylor for her assistance in publishing this book.

Last but not least, my sincere gratitude goes to my family for their continuous support and encouragement.

Acronyms

A word means just what I choose it to mean – neither more nor less.

Lewis Carroll

ACR	Absolute category rating
ANSI	American National Standards Institute
ATM	Asynchronous transfer mode
CIE	Commission Internationale de l'Eclairage
cpd	Cycles per degree
CRT	Cathode ray tube
CSF	Contrast sensitivity function
dB	Decibel
DCR	Degradation category rating
DCT	Discrete cosine transform
DMOS	Differential mean opinion score
DSCQS	Double stimulus continuous quality scale
DSIS	Double stimulus impairment scale
DVD	Digital versatile disk
DWT	Discrete wavelete transform
EBU	European Broadcasting Union
FIR	Finite impulse response
HDTV	High-definition television
HLS	Hue, lightness, saturation
HSV	Hue, saturation, value
HVS	Human visual system
IEC	International Electrotechnical Commission
IIR	Infinite impulse response
ISO	International Organization for Standardization
ITU	International Telecommunication Union

JND	Just noticeable difference
JPEG	Joint Picture Experts Group
kb/s	Kilobit per second
LGN	Lateral geniculate nucleus
Mb/s	Megabit per second
MC	Motion compensation
MOS	Mean opinion score
MPEG	Moving Picture Experts Group
MSE	Mean squared error
MSSG	MPEG Software Simulation Group
NTSC	National Television Systems Committee
NVFM	Normalization video fidelity metric
PAL	Phase Alternating Line
PDM	Perceptual distortion metric
PBDM	Perceptual blocking distortion metric
PSNR	Peak signal-to-noise ratio
RGB	Red, green, blue
RMSE	Root mean squared error
SID	Society for Information Display
SSCQE	Single stimulus continuous quality evaluation
SNR	Signal-to-noise ratio
TCP/IP	Transmission control protocol/internet protocol
VCD	Video compact disk
VHS	Video home system
VQEG	Video Quality Experts Group

1

Introduction

'Where shall I begin, please your Majesty?' he asked.
'Begin at the beginning,' the King said, gravely,
'and go on till you come to the end: then stop.'

Lewis Carroll

1.1 MOTIVATION

Humans are highly visual creatures. Evolution has invested a large part of our neurological resources in visual perception. We are experts at grasping visual environments in a fraction of a second and rely on visual information for many of our day-to-day activities. It is not surprising that, as our world is becoming more digital every day, digital images and digital video are becoming ubiquitous.

In light of this development, optimizing the performance of digital imaging systems with respect to the capture, display, storage and transmission of visual information is one of the most important challenges in this domain. Video compression schemes should reduce the visibility of the introduced artifacts, watermarking schemes should hide information more effectively in images, printers should use the best half-toning patterns, and so on. In all these applications, the limitations of the human visual system (HVS) can be exploited to maximize the visual quality of the output. To do this, it is necessary to build computational models of the HVS and integrate them in tools for perceptual quality assessment.

Digital Video Quality - Vision Models and Metrics   Stefan Winkler
© 2005 John Wiley & Sons, Ltd   ISBN: 0-470-02404-6

The need for accurate vision models and quality metrics has been increasing as the borderline between analog and digital processing of visual information is moving closer to the consumer. This is particularly evident in the field of television. While traditional analog systems still represent the majority of television sets today, production studios, broadcasters and network providers have been installing digital video equipment at an ever-increasing rate. Digital satellite and cable services have been available for quite some time, and terrestrial digital TV broadcast has been introduced in a number of locations around the world. A similar development can be observed in photography, where digital cameras have become hugely popular.

The advent of digital imaging systems has exposed the limitations of the techniques traditionally used for quality assessment and control. For conventional analog systems there are well-established performance standards. They rely on special test signals and measurement procedures to determine signal parameters that can be related to perceived quality with relatively high accuracy. While these parameters are still useful today, their connection with perceived quality has become much more tenuous. Because of compression, digital imaging systems exhibit artifacts that are fundamentally different from analog systems. The amount and visibility of these distortions strongly depend on the actual image content. Therefore, traditional measurements are inadequate for the evaluation of these artifacts.

Given these limitations, researchers have had to resort to subjective viewing experiments in order to obtain reliable ratings for the quality of digital images or video. While these tests are the best way to measure 'true' perceived quality, they are complex, time-consuming and consequently expensive. Hence, they are often impractical or not feasible at all, for example when real-time online quality monitoring of several video channels is desired.

Looking for faster alternatives, the designers of digital imaging systems have turned to simple error measures such as mean squared error (MSE) or peak signal-to-noise ratio (PSNR), suggesting that they would be equally valid. However, these simple measures operate solely on a pixel-by-pixel basis and neglect the important influence of image content and viewing conditions on the actual visibility of artifacts. Therefore, their predictions often do not agree well with actual perceived quality.

These problems have prompted the intensified study of vision models and visual quality metrics in recent years. Approaches based on HVS-models are slowly replacing classical schemes, in which the quality metric consists of an MSE- or PSNR-measure. The quality improvement that can be achieved using an HVS-based approach instead is significant and applies to a large variety of image processing applications. However, the human visual system is extremely complex, and many of its properties are not well understood even today. Significant advancements of the current state of the art will require an in-depth understanding of human vision for the design of reliable models.

The purpose of this book is to provide an introduction to vision modeling in the framework of video quality assessment. We will discuss the design of models and metrics and show examples of their utilization. The models presented are quite general and may be useful in a variety of image and video processing applications.

1.2 OUTLINE

Chapter 2 gives an overview of the human visual system. It looks at the anatomy and physiology of its components, explaining the processing of visual information in the brain together with the resulting perceptual phenomena.

Chapter 3 outlines the main aspects of visual quality with a special focus on digital video. It briefly introduces video coding techniques and explores the effects that lossy compression or transmission errors have on quality. We take a closer look at factors that can influence subjective quality and describe procedures for its measurement. Then we review the history and state of the art of video quality metrics and discuss the evaluation of their prediction performance.

Chapter 4 presents tools for vision modeling and quality measurement. The first is a unique measure of isotropic local contrast based on analytic directional filters. It agrees well with perceived contrast and is used later in conjunction with quality assessment. The second tool is a perceptual distortion metric (PDM) for the evaluation of video quality. It is based on a model of the human visual system that takes into account color perception, the multi-channel architecture of temporal and spatial mechanisms, spatio-temporal contrast sensitivity, pattern masking and channel interactions.

Chapter 5 is devoted to the evaluation of the prediction performance of the PDM as well as a comparison with competing metrics. This is achieved with the help of extensive data from subjective experiments. Furthermore, the design choices for the different components of the PDM are analyzed with respect to their influence on prediction performance.

Chapter 6 investigates a number of extensions of the perceptual distortion metric. These include modifications of the PDM for the prediction of perceived blocking distortions and for the support of object segmentation. Furthermore, attributes of image appeal are integrated in the PDM in the form of sharpness and colorfulness ratings derived from the video. Additional data from subjective experiments are used in each case for the evaluation of prediction performance.

Finally, Chapter 7 concludes the book with an outlook on promising developments in the field of video quality assessment.

2
Vision

Seeing is believing.

English proverb

Vision is the most essential of our senses; 80–90% of all neurons in the human brain are estimated to be involved in visual perception (Young, 1991). This is already an indication of the enormous complexity of the human visual system. The discussions in this chapter are necessarily limited in scope and focus mostly on aspects relevant to image and video processing. For a more detailed overview of vision, the reader is referred to the abundant literature, e.g. the excellent book by Wandell (1995).

The human visual system can be subdivided into two major components: the eyes, which capture light and convert it into signals that can be understood by the nervous system, and the visual pathways in the brain, along which these signals are transmitted and processed. This chapter discusses the anatomy and physiology of these components as well as a number of phenomena of visual perception that are of particular relevance to the models and metrics discussed in this book.

2.1 EYE

2.1.1 Physical Principles

From an optical point of view, the eye is the equivalent of a photographic camera. It comprises a system of lenses and a variable aperture to focus

Digital Video Quality - Vision Models and Metrics   Stefan Winkler
© 2005 John Wiley & Sons, Ltd   ISBN: 0-470-02404-6

images on the light-sensitive retina. This section summarizes the basics of the optical principles of image formation (Bass *et al.*, 1995; Hecht, 1997).

The optics of the eye rely on the physical principles of *refraction*. Refraction is the bending of light rays at the angulated interface of two transparent media with different refractive indices. The refractive index n of a material is the ratio of the speed of light in vacuum c_0 to the speed of light in this material c: $n = c_0/c$. The degree of refraction depends on the ratio of the refractive indices of the two media as well as the angle ϕ between the incident light ray and the interface normal: $n_1 \sin \phi_1 = n_2 \sin \phi_2$. This is known as *Snell's law*.

Lenses exploit refraction to converge or diverge light, depending on their shape. Parallel rays of light are bent outwards when passing through a concave lens and inwards when passing through a convex lens. These focusing properties of a convex lens can be used for image formation. Due to the nature of the projection, the image produced by the lens is reversed, i.e. rotated 180° about the optical axis.

Objects at different distances from a convex lens are focused at different distances behind the lens. In a first approximation, this is described by the *Gaussian lens formula*:

$$\frac{1}{d_s} + \frac{1}{d_i} = \frac{1}{f}, \tag{2.1}$$

where d_s is the distance between the source and the lens, d_i is the distance between the image and the lens, and f is the *focal length* of the lens. An infinitely distant object is focused at focal length, $d_i = f$. The reciprocal of the focal length is a measure of the *optical power* of a lens, i.e. how strongly incoming rays are bent. The optical power is defined as $1m/f$ and is specified in *diopters*.

A variable aperture is added to most optical imaging systems in order to adapt to different light levels. Apart from limiting the amount of light entering the system, the aperture size also influences the *depth of field*, i.e. the range of distances over which objects will appear in focus on the imaging plane. A small aperture produces images with a large depth of field, and vice versa.

Another side-effect of an aperture is *diffraction*. Diffraction is the scattering of light that occurs when the extent of a light wave is limited. The result is a blurred image. The amount of blurring depends on the dimensions of the aperture in relation to the wavelength of the light.

A final note regarding notation: distance-independent specifications of images are often used in optics. The size is measured in terms of visual angle

$\alpha = \mathrm{atan}(s/2D)$ covered by an image of size s at distance D. Accordingly, spatial frequencies are measured in cycles per degree (cpd) of visual angle.

2.1.2 Optics of the Eye

Making general statements about the eye's optical characteristics is complicated by the fact that there are considerable variations between individuals. Furthermore, its components undergo continuous changes throughout life. Therefore, the figures given in the following should be considered approximate.

The optical system of the human eye is composed of the cornea, the aqueous humor, the lens, and the vitreous humor, as illustrated in Figure 2.1. The refractive indices of these four components are 1.38, 1.33, 1.40, and

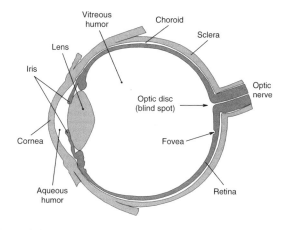

Figure 2.1 The human eye (transverse section of the left eye).

1.34, respectively (Guyton, 1991). The total optical power of the eye is approximately 60 diopters. Most of it is provided by the air–cornea transition, because this is where the largest difference in refractive indices occurs (the refractive index of air is close to 1). The lens itself provides only a third of the total refractive power due to the optically similar characteristics of the surrounding elements.

The importance of the lens is that its curvature and thus its optical power can be voluntarily increased by contracting muscles attached to it. This process is called *accommodation*. Accommodation is essential to bring objects at different distances into focus on the retina. In young children, the optical power of the lens can be increased from 20 to 34 diopters.

However, accommodation ability decreases gradually with age until it is lost almost completely, a condition known as *presbyopia*.

Just before entering the lens, the light passes the *pupil*, the eye's aperture. The pupil is the circular opening inside the *iris*, a set of muscles that control its size and thus the amount of light entering the eye depending on the exterior light levels. Incidentally, the pigmentation of the iris is also responsible for the color of our eyes. The diameter of the pupillary aperture can be varied between 1.5 and 8 mm, corresponding to a 30-fold change of the quantity of light entering the eye. The pupil is thus one of the mechanisms of the human visual system for light adaptation (cf. section 2.4.1).

2.1.3 Optical Quality

The physical principles described in section 2.1.1 pertain to an ideal optical system, whose resolution is only limited by diffraction. While the parameters of an individual healthy eye are usually correlated in such a way that the eye can produce a sharp image of a distant object on the retina (Charman, 1995), imperfections in the lens system can introduce additional distortions that affect image quality. In general, the optical quality of the eye deteriorates with increasing distance from the optical axis (Liang and Westheimer, 1995). This is not a severe problem, however, because visual acuity also decreases there, as will be discussed in section 2.2.

To determine the optical quality of the eye, the reflection of a visual stimulus projected onto the retina can be measured (Campbell and Gubisch, 1966).[†] The retinal image turns out to be a distorted version of the input, the most noticeable distortion being blur. To quantify the amount of blurring, a point or a thin line is used as the input image, and the resulting retinal image is called the *point spread function* or *line spread function* of the eye; its Fourier transform is the *modulation transfer function*. A simple approximation of the foveal point spread function of the human eye according to Westheimer (1986) is shown in Figure 2.2 for a pupil diameter of 3 mm. The amount of blurring depends on the pupil size: for small pupil diameters up to 3–4 mm, the optical blurring is close to the diffraction limit; as the pupil diameter increases (for lower ambient light levels), the width of the point spread function increases as well, because the distortions due to cornea and lens imperfections become large compared to diffraction effects (Campbell and Gubisch, 1966; Rovamo *et al.*, 1998). The pupil size also influences the depth of field, as mentioned before.

[†]An alternative method to determine the optical quality of the eye is based on interferometric measurements. A comparison of these two methods is given by Williams *et al.* (1994).

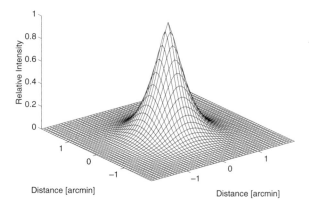

Figure 2.2 Point spread function of the human eye as a function of visual angle (Westheimer, 1986).

Because the cornea is not perfectly symmetric, the optical properties of the eye are orientation-dependent. Therefore it is impossible to perfectly focus stimuli of all orientations simultaneously, a condition known as *astigmatism*. This results in a point spread function that is not circularly symmetric. Astigmatism can be severe enough to interfere with perception, in which case it has to be corrected by compensatory glasses.

The properties of the eye's optics, most importantly the refractive indices of the optical elements, also vary with wavelength. This means that it is impossible to focus all wavelengths simultaneously, an effect known as *chromatic aberration*. The point spread function thus changes with wavelength. Chromatic aberration can be quantified by determining the modulation transfer function of the human eye for different wavelengths. This is shown in Figure 2.3 for a human eye model with a pupil diameter of 3 mm and in focus at 580 nm (Marimont and Wandell, 1994).

It is evident that the retinal image contains only poor spatial detail at wavelengths far from the in-focus wavelength (note the sharp cutoff going down to a few cycles per degree at short wavelengths). This tendency towards monochromaticity becomes even more pronounced with increasing pupil aperture.

2.1.4 Eye Movements

The eye is attached to the head by three pairs of muscles that provide for rotation around its three axes. Several different types of eye movements can be distinguished (Carpenter, 1988). Fixation movements are perhaps the most

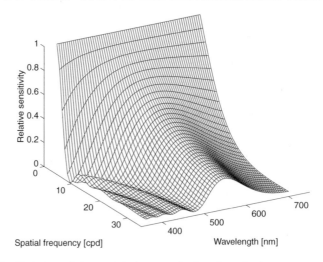

Relative sensitivity

Spatial frequency [cpd]

Wavelength [nm]

Figure 2.3 Variation of the modulation transfer function of a human eye model with wavelength (Marimont and Wandell, 1994).

important. The *voluntary fixation mechanism* allows us to direct the eyes towards an object of interest. This is achieved by means of *saccades*, high-speed movements steering the eyes to the new position. Saccades occur at a rate of 2–3 per second and are also used to scan a scene by fixating on one highlight after the other. One is unaware of these movements because the visual image is suppressed during saccades. The *involuntary fixation mechanism* locks the eyes on the object of interest once it has been found. It involves so-called micro-saccades that counter the tremor and slow drift of the eye muscles. As soon as the target leaves the fovea, it is re-centered with the help of these small flicking movements. The same mechanism also compensates for head movements or vibrations.

Additionally, the eyes can track an object that is moving across the scene. These so-called *pursuit movements* can adapt to object trajectories with great accuracy. Smooth pursuit works well even for high velocities, but it is impeded by large accelerations and unpredictable motion (Eckert and Buchsbaum, 1993; Hearty, 1993).

2.2 RETINA

The optics of the eye project images of the outside world onto the *retina*, the neural tissue at the back of the eye. The functional components of the retina

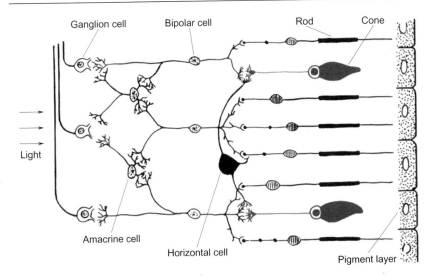

Figure 2.4 Anatomy of the retina.

are illustrated in Figure 2.4. Light entering the retina has to traverse several layers of neurons before it reaches the light-sensitive layer of photoreceptors and is finally absorbed in the pigment layer. The anatomy and physiology of the photoreceptors and the retinal neurons is discussed in more detail here.

2.2.1 Photoreceptors

The photoreceptors are specialized neurons that make use of light-sensitive photochemicals to convert the incident light energy into signals that can be interpreted by the brain. There are two different types of photoreceptors, namely *rods* and *cones*. The names are derived from the physical appearance of their light-sensitive outer segments. Rods are responsible for *scotopic* vision at low light levels, while cones are responsible for *photopic* vision at high light levels.

Rods are very sensitive light detectors. With the help of the photochemical rhodopsin they can generate a photocurrent response from the absorption of only a single photon (Hecht *et al.*, 1942; Baylor, 1987). However, visual acuity under scotopic conditions is poor, even though rods sample the retina very finely. This is due to the fact that signals from many rods converge onto a single neuron, which improves sensitivity but reduces resolution.

The opposite is true for the cones. Several neurons encode the signal from each cone, which already suggests that cones are important components of visual processing. There are three different types of cones, which can be classified according to the spectral sensitivity of their photochemicals. These three types are referred to as L-cones, M-cones, and S-cones, according to their sensitivity to long, medium, and short wavelengths, respectively.[†] They form the basis of color perception. Recent estimates of the absorption spectra of the three cone types are shown in Figure 2.5.

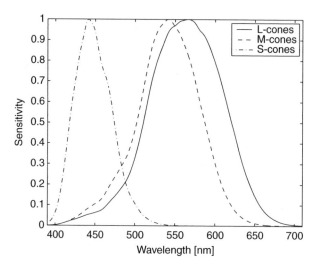

Figure 2.5 Normalized absorption spectra of the three cone types: L-cones (solid), M-cones (dashed), and S-cones (dot-dashed) (Stockman *et al.*, 1999; Stockman and Sharpe, 2000).

The peak sensitivities occur around 440 nm, 540 nm, and 570 nm. As can be seen, the absorption spectra of the L- and M-cones are very similar, whereas the S-cones exhibit a significantly different sensitivity curve. The overlap of the spectra is essential to fine color discrimination. Color perception is discussed in more detail in section 2.5.

There are approximately 5 million cones and 100 million rods in each eye. Their density varies greatly across the retina, as is evident from Figure 2.6 (Curcio *et al.*, 1990). There is also a large variability between individuals. Cones are concentrated in the *fovea*, a small area near the center of the retina, where they can reach a peak density of up to 300 000/mm^2 (Ahnelt, 1998). Throughout the retina, L- and M-cones are in the majority; S-cones are much

---

[†]Sometimes they are also referred to as red, green, and blue cones, respectively.

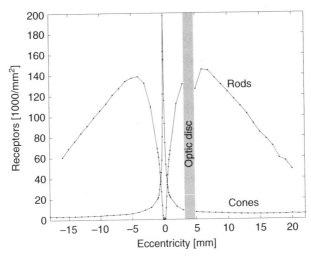

Figure 2.6 The distribution of photoreceptors on the retina. Cones are concentrated in the fovea at the center of the retina, whereas rods dominate in the periphery. The gap around 4 mm eccentricity represents the optic disc, where no receptors are present (Adapted from C. A. Curcio *et al.*, (1990), Human photoreceptor topography, *Journal of Comparative Neurology* **292**: 497–523. Copyright © 1990 John Wiley & Sons. The material is used by permission of Wiley-Liss, Inc., a Subsidiary of John Wiley & Sons, Inc.).

more sparse and account for less than 10% of the total number of cones (Curcio *et al.*, 1991). Rods dominate outside of the fovea, which explains why it is easier to see very dim objects (e.g. stars) when they are in the peripheral field of vision than when looking straight at them. The central fovea contains no rods at all. The highest rod densities (up to $200\,000/mm^2$) are found along an elliptical ring near the eccentricity of the optic disc. The *blind spot* around the optic disc, where the optic nerve exits the eye, is completely void of photoreceptors.

The spatial sampling of the retina by the photoreceptors is illustrated in Figure 2.7. In the fovea the cones are tightly packed and form a very regular hexagonal sampling array. In the periphery the sampling grid becomes more irregular; the separation between the cones grows, and rods fill in the spaces. Also note the size differences: the cones in the fovea have a diameter of 1–3 µm; in the periphery, their diameter increases to 5–10 µm. The diameter of the rods varies between 1 and 5 µm.

The size and spacing of the photoreceptors determine the maximum spatial resolution of the human visual system. Assuming an optical power of 60 diopters and thus a focal length of approximately 17 mm for the eye,

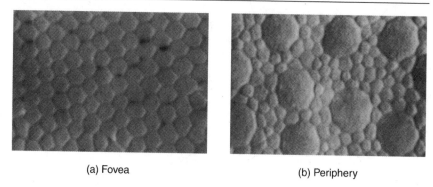

(a) Fovea   (b) Periphery

Figure 2.7 The photoreceptor mosaic on the retina. In the fovea (a) the cones are densely packed on a hexagonal sampling array. In the periphery (b) their size and separation grows, and rods fill in the spaces. Each image shows an area of $35 \times 25 \ \mu m^2$ (Adapted from C. A. Curcio *et al.*, (1990), Human photoreceptor topography, *Journal of Comparative Neurology* 292: 497–523. Copyright © 1990 John Wiley & Sons. The material is used by permission of Wiley-Liss, Inc., a Subsidiary of John Wiley & Sons, Inc.).

distances on the retina can be expressed in terms of visual angle using simple trigonometry. The entire fovea covers approximately 2° of visual angle. The L- and M-cones in the fovea are spaced approximately 2.5 μm apart, which corresponds to 30 arc seconds of visual angle. The maximum resolution of around 60 cpd attained here is high enough to capture all of the spatial variation after the blurring by the eye's optics. S-cones are spaced approximately 50 μm or 10 minutes of arc apart on average, resulting in a maximum resolution of only 3 cpd (Curcio *et al.*, 1991). This is consistent with the strong defocus of short-wavelength light due to the axial chromatic aberration of the eye's optics (see Figure 2.3). Thus the properties of different components of the visual system fit together nicely, as can be expected from an evolutionary system. The optics of the eye set limits on the maximum visual acuity, and the arrangements of the mosaic of the S-cones as well as the L- and M-cones can be understood as a consequence of the optical limitations (and vice versa).

2.2.2 Retinal Neurons

The retinal neurons process the photoreceptor signals. The anatomical connections and neural specializations within the retina combine to communicate different types of information about the visual input to the brain. As shown in Figure 2.4, a variety of different neurons can be distinguished in the retina (Young, 1991):

- *Horizontal cells* connect the synaptic nodes of neighboring rods and cones. They have an inhibitory effect on bipolar cells.
- *Bipolar cells* connect horizontal cells, rods and cones with ganglion cells. Bipolar cells can have either excitatory or inhibitory outputs.
- *Amacrine cells* transmit signals from bipolar cells to ganglion cells or laterally between different neurons. About 30 types of amacrine cells with different functions have been identified.
- *Ganglion cells* collect information from bipolar and amacrine cells. There are about 1.6 million ganglion cells in the retina. Their axons form the *optic nerve* that leaves the eye through the optic disc and carries the output signal of the retina to other processing centers in the brain (see section 2.3).

The interconnections between these cells give rise to an important concept in visual perception, the *receptive field*. The visual receptive field of a neuron is defined as the retinal area in which light influences the neuron's response. It is not limited to cells in the retina; many neurons in later stages of the visual pathways can also be described by means of their receptive fields (see section 2.3.2).

The ganglion cells in the retina have a characteristic center–surround receptive field, which is nearly circularly symmetric, as shown in Figure 2.8

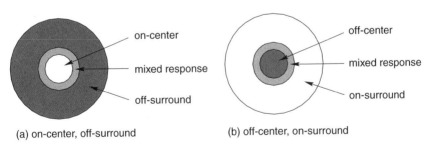

(a) on-center, off-surround   (b) off-center, on-surround

Figure 2.8   Center–surround organization of the receptive field of retinal ganglion cells. Light falling on the center of a ganglion cell's receptive field may either excite (a) or inhibit (b) the cell. In the surrounding region, light has the opposite effect. Between center and surround, there is a small area with a mixed response.

(Kuffler, 1953). Light falling directly on the center of a ganglion cell's receptive field may either excite or inhibit the cell. In the surrounding region, light has the opposite effect. Between center and surround, there is a small area with a mixed response. About half of the retinal ganglion cells have an on-center, off-surround receptive field, i.e. they are excited by light on their center, and the other half have an off-center, on-surround receptive field with the opposite reaction.

This receptive field organization is mainly due to lateral inhibition from horizontal cells. The consequence is that excitatory and inhibitory signals basically neutralize each other when the stimulus is uniform, but when contours or edges come to lie over such a cell's receptive field, its response is amplified. In other words, retinal neurons implement a mechanism of contrast computation. Ganglion cells can be further classified in two main groups (Sekuler and Blake, 1990):

- *P-cells* constitute the large majority (nearly 90%) of ganglion cells. They have very small receptive fields, i.e. they receive inputs only from a small area of the retina (only a single cone in the fovea) and can thus encode fine image details. Furthermore, P-cells encode most of the chromatic information as different P-cells respond to different colors.
- *M-cells* constitute only 5–10% of ganglion cells. At any given eccentricity, their receptive fields are several times larger than those of P-cells. They also have thicker axons, which means that their output signals travel at higher speeds. M-cells respond to motion or small differences in light level, but are insensitive to color. They are responsible for rapidly alerting the visual system to changes in the image.

These two types of ganglion cells represent the origins of two separate visual streams in the brain, the so-called *magnocellular* and *parvocellular pathways* (see section 2.3.1).

As becomes evident from this intricate arrangement of neurons, the retina is much more than a device to convert light to neural signals; the visual information is thoroughly pre-processed here before it is passed on to other parts of the brain.

2.3 VISUAL PATHWAYS

The optic nerve leaves the eye to carry the visual information from the ganglion cells of the retina to various processing centers in the brain. These visual pathways are illustrated in Figure 2.9. The optic nerves from the two eyes meet at the *optic chiasm*, where the fibers are rearranged. All the fibers from the nasal halves of each retina cross to the opposite side, where they join the fibers from the temporal halves of the opposite retinas to form the *optic tracts*. Since the retinal images are reversed by the optics, the left visual field is thus processed in the right hemisphere, and the right visual field is

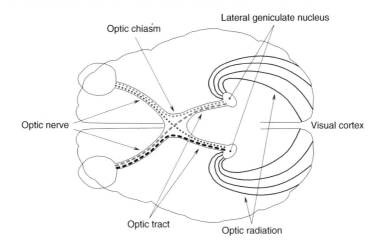

Figure 2.9 Visual pathways in the human brain (transverse section). The signals travel from the eyes through the optic nerves. They meet at the optic chiasm, where the fibers from the nasal halves of each retina cross to the opposite side to join the fibers from the temporal halves of the opposite retinas. From there, the optic tracts lead the signals to the lateral geniculate nuclei and on to the visual cortex.

processed in the left hemisphere. Most of the fibers from each optic tract synapse in the *lateral geniculate nucleus* (see section 2.3.1). From there fibers pass by way of the *optic radiation* to the *visual cortex* (see section 2.3.2). Throughout these visual pathways, the neighborhood relations of the retina are preserved, i.e. the input from a certain small part of the retina is processed in a particular area of the LGN and of the primary visual cortex. This property is known as *retinotopic mapping*.

There are a number of additional destinations for visual information in the brain apart from the major visual pathways listed above. These brain areas are responsible mainly for behavioral or reflex responses. One particular example is the *superior colliculus*, which seems to be involved in controlling eye movements in response to certain stimuli in the periphery.

2.3.1 Lateral Geniculate Nucleus

The lateral geniculate nucleus (LGN) comprises approximately one million neurons in six layers. The two inner layers, the *magnocellular layers*, receive input almost exclusively from M-type ganglion cells. The four outer layers, the *parvocellular layers*, receive input mainly from P-type ganglion cells. As mentioned in section 2.2.2, the M- and P-cells respond to different types of stimuli, namely motion and spatial detail, respectively. This functional specialization continues in the LGN and the visual cortex, which suggests the existence of separate magnocellular and parvocellular pathways in the visual system.

The specialization of cells in the LGN is similar to the ganglion cells in the retina. The cells in the magnocellular layers are effectively color-blind and have larger receptive fields. They respond vigorously to moving contours. The cells in the parvocellular layers have rather small receptive fields and are differentially sensitive to color (De Valois *et al.*, 1958). They are excited if a particular color illuminates the center of their receptive field and inhibited if another color illuminates the surround. Only two color pairings are found, namely red-green and blue-yellow. These *opponent colors* form the basis of color perception in the human visual system and will be discussed in more detail in section 2.5.2.

The LGN serves not only as a relay station for signals from the retina to the visual cortex, but it also controls how much of the information is allowed to pass. This gating operation is controlled by extensive feedback signals from the primary visual cortex as well as input from the reticular activating system in the brain stem, which governs our general level of arousal.

2.3.2 Visual Cortex

The visual cortex is located at the back of the cerebral hemispheres (see section 2.3). It is responsible for all higher-level aspects of vision. The signals from the lateral geniculate nucleus arrive at an area called the *primary visual cortex* (also known as area V1, Brodmann area 17, or striate cortex), which makes up the largest part of the human visual system. In addition to the primary visual cortex, more than 20 other cortical areas receiving strong visual input have been discovered. Little is known about their exact functionalities, however.

There is an enormous variety of cells in the visual cortex. Neurons in the first stage of the primary visual cortex have center–surround receptive fields similar to cells in the retina and in the lateral geniculate nucleus. A recurring property of many cells in the subsequent stages of the visual cortex is their selective sensitivity to certain types of information. A particular cell may respond strongly to patterns of a certain orientation or to motion in a certain direction. Similarly, there are cells tuned to particular frequencies, colors, velocities, etc. This neuronal selectivity is thought to be at the heart of the multi-channel organization of human vision (see section 2.7).

The foundations of our knowledge about cortical receptive fields were laid by Hubel and Wiesel (1959, 1962, 1968, 1977). In their physiological studies of cells in the primary visual cortex, they were able to identify several classes of neurons with different specializations. *Simple cells* behave in an approximately linear fashion, i.e. their responses to complicated shapes can be predicted from their responses to small-spot stimuli. They have receptive fields composed of several parallel elongated excitatory and inhibitory regions, as illustrated in Figure 2.10. In fact, their receptive fields resemble Gabor patterns (Daugman, 1980). Hence, simple cells can be characterized by a particular spatial frequency, orientation, and phase. Serving as an oriented band-pass filter, a simple cell thus responds to a certain range of spatial frequencies and orientations about its center values.

Figure 2.10   Idealized receptive field of a simple cell in the primary visual cortex. Light and dark shades denote excitatory and inhibitory regions, respectively.

Complex cells are the most common cells in the primary visual cortex. Like simple cells, they are also orientation-selective, but their receptive field does not exhibit the on- and off-regions of a simple cell; instead, they respond to a properly oriented stimulus anywhere in their receptive field.

A small percentage of complex cells respond well only when a stimulus (still with the proper orientation) moves across their receptive field in a certain direction. These *direction-selective cells* receive input mainly from the magnocellular pathway and probably play an important role in motion perception. Some cells respond only to oriented stimuli of a certain size. They are referred to as *end-stopped cells*. They are sensitive to corners, curvature or sudden breaks in lines. Both simple and complex cells can also be end-stopped. Furthermore, the primary visual cortex is the first stage in the visual pathways where individual neurons have binocular receptive fields, i.e. they receive inputs from both eyes, thereby forming the basis for stereopsis and depth perception (Hubel, 1995).

2.4 SENSITIVITY TO LIGHT

2.4.1 Light Adaptation

The human visual system is capable of adapting to an enormous range of light intensities. Light adaptation allows us to better discriminate relative luminance variations at every light level. Scotopic and photopic vision together cover 12 orders of magnitude in intensity, from a few photons to bright sunlight (Hood and Finkelstein, 1986). However, at any given level of adaptation we can only discriminate within an intensity range of 2–3 orders of magnitude (Rogowitz, 1983).

Three mechanisms for light adaptation can be distinguished in the human visual system (Guyton, 1991):

- The mechanical variation of the pupillary aperture. As discussed in section 2.1.2, this is controlled by the iris. The pupil diameter can be varied between 1.5 and 8 mm, which corresponds to a 30-fold change of the quantity of light entering the eye. This adaptation mechanism responds in a matter of seconds.
- The chemical processes in the photoreceptors. This adaptation mechanism exists in both rods and cones. In bright light, the concentration of photochemicals in the receptors decreases, thereby reducing their sensitivity. On the other hand, when the light intensity is reduced, the production of photochemicals and thus the receptor sensitivity is increased. While this chemical adaptation mechanism is very powerful (it covers 5–6 orders of magnitude), it is rather slow; complete dark adaptation in particular can take up to an hour.
- Adaptation at the neural level. This mechanism involves the neurons in all layers of the retina, which adapt to changing light intensities by increasing or decreasing their signal output accordingly. Neural adaptation is less powerful, but faster than the chemical adaptation in the photoreceptors.

2.4.2 Contrast Sensitivity

The response of the human visual system depends much less on the absolute luminance than on the relation of its local variations to the surrounding luminance. This property is known as the *Weber–Fechner law*. Contrast is a measure of this relative variation of luminance. Mathematically, *Weber contrast* can be expressed as

$$C^W = \frac{\Delta L}{L}. \tag{2.2}$$

This definition is most appropriate for patterns consisting of a single increment or decrement ΔL to an otherwise uniform background luminance.

The threshold contrast, i.e. the minimum contrast necessary for an observer to detect a change in intensity, is shown as a function of background luminance in Figure 2.11. As can be seen, it remains nearly constant over an important range of intensities (from faint lighting to daylight) due to the adaptation capabilities of the human visual system, i.e. the Weber–Fechner law holds in this range. This is indeed the luminance range typically

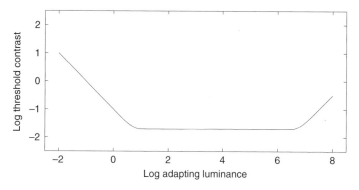

Figure 2.11 Illustration of the Weber–Fechner law. The threshold contrast remains nearly constant over a wide range of intensities.

encountered in most image processing applications. Outside of this range, our intensity discrimination ability deteriorates. Evidently, the Weber–Fechner law is only an approximation of the actual sensory perception, but contrast measures based on this concept are widely used in vision science.

Under optimal conditions, the threshold contrast can be less than 1% (Hood and Finkelstein, 1986). The exact figure depends to a great extent on the stimulus characteristics, most importantly its color as well as its spatial and temporal frequency. Contrast sensitivity functions (CSFs) are generally used to quantify these dependencies. Contrast sensitivity is defined as the inverse of the contrast threshold.

In measurements of the CSF, the contrast of periodic (often sinusoidal) stimuli with varying frequencies is defined as the *Michelson contrast* (Michelson, 1927):

$$C^M = \frac{L_{\max} - L_{\min}}{L_{\max} + L_{\min}}, \qquad (2.3)$$

where $L_{\min}$ and $L_{\max}$ are the luminance extrema of the pattern. Figure 2.12, the so-called Campbell–Robson chart[†] (Campbell and Robson, 1968), demonstrates the shape of the spatial contrast sensitivity function in a very

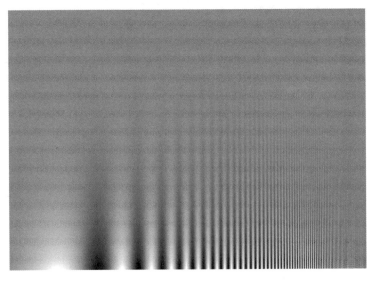

Figure 2.12   Campbell–Robson contrast sensitivity chart (Campbell and Robson, 1968). The spatial CSF appears as the envelope of visibility of the modulated pattern.

intuitive manner. The luminance of pixels is modulated sinusoidally along the horizontal dimension. The frequency of modulation increases exponentially from left to right, while the contrast decreases exponentially from 100% to about 0.5% from bottom to top. The minimum and maximum luminance remain constant along any given horizontal line through the image. Therefore, if the detection of contrast were dictated solely by

[†]Several renditions of this chart are available at http://www.bpe.es.osaka-u.ac.jp/ohzawa-lab/izumi/CSF/A_JG_RobsonCSFchart.html

image contrast, the alternating bright and dark bars should appear to have equal height everywhere in the image. However, the bars appear taller in the middle of the image than at the sides. This inverted U-shape of the envelope of visibility is the spatial contrast sensitivity function for sinusoidal stimuli. The location of its peak depends on the viewing distance.

Spatio-temporal CSF approximations are shown in Figure 2.13. Achromatic contrast sensitivity is generally higher than chromatic, especially for high spatio-temporal frequencies. The chromatic CSFs for red-green and blue-yellow stimuli are very similar in shape; however, the blue-yellow sensitivity is somewhat lower overall, and its high-frequency decline sets in earlier. Hence, the full range of colors is perceived only at low frequencies. As spatio-temporal frequencies increase, blue-yellow sensitivity declines first. At even higher frequencies, red-green sensitivity diminishes as well, and perception becomes achromatic. On the other hand, achromatic sensitivity decreases at low spatio-temporal frequencies (albeit to a lesser extent), whereas chromatic sensitivity does not. However, this apparent attenuation of sensitivity towards low frequencies may be attributed to implicit masking, i.e. masking by the spectrum of the window within which the test gratings are presented (Yang and Makous, 1997).

There has been some debate about the space–time separability of the spatio-temporal CSF. This property is of interest in vision modeling because a CSF that could be expressed as a product of spatial and temporal components would simplify modeling. Early studies concluded that the spatio-temporal CSF was not space–time separable at lower frequencies (Robson, 1966; Koenderink and van Doorn, 1979). Kelly (1979a) measured contrast sensitivity under stabilized conditions (i.e. the stimuli were stabilized on the retina by compensating for the observers' eye movements). Kelly (1979b) fit an analytic function to his measurements, which yields a very close approximation of the spatio-temporal CSF for counterphase flicker. Burbeck and Kelly (1980) found that this CSF can be approximated by linear combinations of two space–time separable components termed excitatory and inhibitory CSFs. The same holds for the chromatic CSF (Kelly, 1983).

Yang and Makous (1994) measured the spatio-temporal CSF for both in-phase and conventional counterphase modulation. Their results suggest that the underlying filters are indeed spatio-temporally separable and have the shape of low-pass exponentials. The spatio-temporal interactions observed for counterphase modulation may be explained as a product of masking by the zero-frequency component of the gratings.

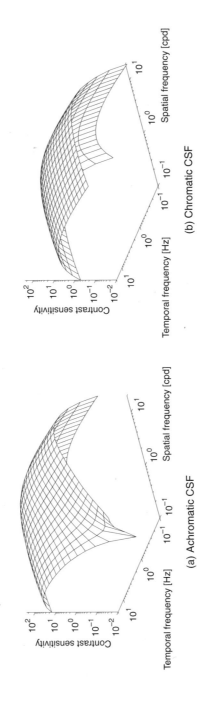

Figure 2.13 Approximations of achromatic (a) and chromatic (b) spatio-temporal contrast sensitivity functions (Kelly, 1979b; Burbeck and Kelly, 1980; Kelly, 1983).

2.5 COLOR PERCEPTION

In its most general form, light can be described by its spectral power distribution. The human visual system, however, uses a much more compact representation of color, which will be discussed in this section.

2.5.1 Color Matching

Color perception can be studied by the *color-matching experiment* (Brainard, 1995). It is the foundation of color science and has many applications. In the color-matching experiment, the observer views a bipartite field, half of which is illuminated by a test light, the other half by an additive mixture of a certain number of primary lights. The observer is asked to adjust the intensities of the primary lights to match the appearance of the test light.

It is not a priori clear that it will be possible for the observer to make a match when the number of primaries is small. In general, however, observers are able to establish a match using only three primary lights. This is referred to as the *trichromacy* of human color vision.[†] Trichromacy implies that there exist lights with different spectral power distributions that cannot be distinguished by a human observer. Such physically different lights that produce identical color appearance are called *metamers*.

As was first established by Grassmann (1853), photopic color matching satisfies homogeneity and superposition and can thus be analyzed using linear systems theory. Assume the test light is known by N samples of its spectral distribution, expressed as vector $\mathbf{x}$. The color-matching experiment can then be described by

$$\mathbf{t} = \mathbf{Cx}, \tag{2.4}$$

where $\mathbf{t}$ is a three-dimensional vector whose coefficients are the intensities of the three primary lights found by the observer to visually match $\mathbf{x}$. They are also referred to as the *tristimulus coordinates* of the test light. The rows of matrix $\mathbf{C}$ are made up of N samples of the so-called *color-matching functions* of the three primaries; they do not represent spectral power distributions, however.

[†]There are certain qualifications to the empirical generalization that three primaries are sufficient to match any test light. The primary lights must be chosen so that they are visually independent, i.e. no additive mixture of any two of the primary lights should be a match to the third. Also, 'negative' intensities of a primary must be allowed, which is just a mathematical convention of saying that a primary can be added to the test light instead of to the other primaries.

The mechanistic explanation of the color-matching experiment is that two lights match if they produce the same absorption rates in the L-, M-, and S-cones. If the spectral sensitivities of the three cone types (see Figure 2.5) are represented by the rows of a matrix $\mathbf{R}$, the absorption rates of the cones in response to a test light with spectral power distribution $\mathbf{x}$ are given by $\mathbf{r} = \mathbf{Rx}$. To relate these cone absorption rates to the tristimulus coordinates of the test light, we perform a color-matching experiment with primaries $\mathbf{P}$, whose columns contain N samples of the spectral power distribution of the three primaries. It turns out that the cone absorption rates $\mathbf{r}$ are related to the tristimulus coordinates $\mathbf{t}$ of the test light by a linear transformation,

$$\mathbf{r} = \mathbf{Mt}, \qquad\qquad (2.5)$$

where $\mathbf{M} = \mathbf{R_P}$ is a 3×3 matrix. This also implies that the color-matching functions are determined by the cone sensitivities up to a linear transformation, which was first verified empirically by Baylor (1987). The spectral sensitivities of the three cone types thus provide a satisfactory explanation of the color-matching experiment.

2.5.2 Opponent Colors

Hering (1878) was the first to point out that some pairs of hues can coexist in a single color sensation (e.g. a reddish yellow is perceived as orange), while others cannot (we never perceive a reddish green, for instance). This led him to the conclusion that the sensations of red and green as well as blue and yellow are encoded as color difference signals in separate visual pathways, which is commonly referred to as the theory of *opponent colors*.

Empirical evidence in support of this theory came from a behavioral experiment designed to quantify opponent colors, the so-called *hue-cancellation experiment* (Jameson and Hurvich, 1955; Hurvich and Jameson, 1957). In the hue-cancellation experiment, observers are able to cancel, for example, the reddish appearance of a test light by adding certain amounts of green light. Thus the red-green or blue-yellow appearance of monochromatic lights can be measured.

Physiological experiments revealed the existence of opponent signals in the visual pathways (Svaetichin, 1956; De Valois *et al.*, 1958). They demonstrated that cones may have an excitatory or an inhibitory effect on ganglion cells in the retina and on cells in the lateral geniculate nucleus. Depending on the cone types, certain excitation/inhibition pairings occur much more often than others: neurons excited by 'red' L-cones are usually inhibited by 'green' M-cones, and neurons excited by 'blue' S-cones are often inhibited by a combination of L- and M-cones. Hence, the receptive fields of these neurons suggest a connection between neural signals and perceptual opponent colors.

The decorrelation of cone signals achieved by the opponent-signal representation of color information in the human visual system improves the coding efficiency of the visual pathways. In fact, this representation may be the result of the properties of natural spectra (Lee *et al.*, 2002). The precise opponent-color directions are still subject to debate, however. As an example, the spectral sensitivities of an opponent color space derived by Poirson and Wandell (1993) are shown in Figure 2.14. The principal

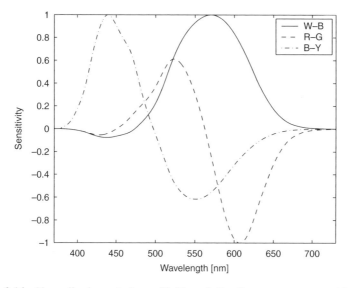

Figure 2.14 Normalized spectral sensitivities of the three components white-black (solid), red-green (dashed), and blue-yellow (dot-dashed) of the opponent color space derived by Poirson and Wandell (1993).

components are white-black (W-B), red-green (R-G) and blue-yellow (B-Y) differences. As can be seen, the W-B channel, which encodes luminance information, is determined mainly by medium to long wavelengths. The R-G channel discriminates between medium and long wavelengths, while the B-Y channel discriminates between short and medium wavelengths.

2.6 MASKING AND ADAPTATION

2.6.1 Spatial Masking

Masking and adaptation are very important phenomena in vision in general and in image processing in particular as they describe interactions between stimuli. Results from masking and adaptation experiments were also the major motivation for developing a multi-channel theory of vision (see section 2.7).

Masking occurs when a stimulus that is visible by itself cannot be detected due to the presence of another. Spatial masking effects are usually quantified by measuring the detection threshold for a target stimulus when it is super-imposed on a masker with varying contrast (Legge and Foley, 1980).

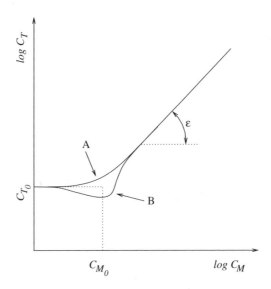

Figure 2.15 Illustration of typical masking curves. For stimuli with different characteristics, masking is the dominant effect (case A). Facilitation occurs for stimuli with similar characteristics (case B).

Figure 2.15 shows an example of curves approximating the data typically resulting from such experiments. The horizontal axis shows the log of the masker contrast C_M, and the vertical axis the log of the target contrast C_T at detection threshold. The detection threshold for the target stimulus without any masker is indicated by C_{T_0}. For contrast values of the masker larger than C_{M_0}, the detection threshold grows with increasing masker contrast.

Two cases can be distinguished in Figure 2.15. In case A, there is a gradual transition from the threshold range to the masking range. Typically this occurs when masker and target have different characteristics. For case B, the detection threshold for the target *decreases* when the masker contrast is close to C_{M_0}, which implies that the target is easier to perceive due to the presence of the masker in this contrast range. This effect is known as *facilitation* and occurs mainly when target and masker have very similar properties.

Masking is strongest when the interacting stimuli have similar characteristics, i.e. similar frequencies, orientations, colors, etc. Masking also occurs between stimuli of different orientation (Foley, 1994) between stimuli of different spatial frequency (Foley and Yang, 1991), and between chromatic and achromatic stimuli (Switkes *et al.*, 1988; Cole *et al.*, 1990; Losada and Mullen, 1994), although it is generally weaker.

Within the framework of image processing it is helpful to think of the distortion or coding noise being masked (or facilitated) by the original image or sequence acting as background. Spatial masking explains why similar artifacts are disturbing in certain regions of an image while they are hardly noticeable elsewhere, as demonstrated in Figure 2.16. In this case, however,

Figure 2.16 Demonstration of masking. Starting from the original image on the left, the same rectangular noise patch was added to regions at the top (center image) and at the bottom (right image). The noise is clearly visible in the sky, whereas it is much harder to see on the rocks and in the water due to the strong masking by these textured regions.

the stimuli are much more complex than those typically used in visual experiments. Because the observer is not familiar with the patterns, uncertainty effects become more important, and masking can be much larger. To account for these effects, a number of different masking mechanisms have been proposed depending on the nature of the masker (Klein *et al.*, 1997; Watson *et al.*, 1997).

2.6.2 Temporal Masking

Temporal masking is an elevation of visibility thresholds due to temporal discontinuities in intensity, for example scene cuts. Within the framework of television, it was first studied by Seyler and Budrikis (1959, 1965), who concluded that the threshold elevation may last up to a few hundred milliseconds after a transition from dark to bright or from bright to dark. More recently, Tam *et al.* (1995) investigated the visibility of MPEG-2 coding artifacts after a scene cut and found significant visual masking effects only in the first subsequent frame. Carney *et al.* (1996) noticed a strong dependence on stimulus polarity, with the masking effect being much more pronounced when target and masker match in polarity. They also found masking to be greatest for local spatial configurations.

Interestingly, temporal masking can occur not only after a discontinuity ('forward masking'), but also before (Breitmeyer and Ogmen, 2000). This 'backward masking' may be explained as the result of the variation in the latency of the neural signals in the visual system as a function of their intensity (Ahumada *et al.* 1998). The opposite of temporal masking, temporal facilitation, can occur at low-contrast discontinuities (Girod, 1989).

2.6.3 Pattern Adaptation

Pattern adaptation adjusts the sensitivity of the visual system in response to the prevalent stimulation patterns. For example, adaptation to patterns of a certain frequency can lead to a noticeable decrease of contrast sensitivity around this frequency (Blakemore and Campbell, 1969; Greenlee and Thomas, 1992; Wilson and Humanski, 1993; Snowden and Hammett, 1996).

An interesting study in this respect was carried out by Webster and Miyahara (1997). They used natural images of outdoor scenes (both distant views and close-ups) as adapting stimuli. It was found that exposure to such stimuli induces pronounced changes in contrast sensitivity. The effects can be characterized by selective losses in sensitivity at lower to medium spatial frequencies. This is consistent with the characteristic amplitude spectra of natural images, which decrease with frequency approximately as $1/f$.

Likewise, Webster and Mollon (1997) examined how color sensitivity and appearance might be influenced by adaptation to the color distributions of images. They found that natural scenes exhibit a limited range of chromatic distributions, so that the range of adaptation states is normally limited as well. However, the variability is large enough for different adaptation effects to occur for individual scenes or for different viewing conditions.

2.7 MULTI-CHANNEL ORGANIZATION

Electrophysiological measurements of the receptive fields of neurons in the lateral geniculate nucleus and in the primary visual cortex (see section 2.3.2) revealed that many of these cells are tuned to certain types of visual information such as color, frequency, and orientation. Data from experiments on pattern discrimination, masking, and adaptation (see section 2.6) yielded further evidence that these stimulus characteristics are processed in different channels in the human visual system. This empirical evidence motivated the multi-channel theory of human vision (Braddick *et al.*, 1978). While this theory is challenged by certain other experiments (Wandell, 1995), it provides an important framework for understanding and modeling pattern sensitivity.

2.7.1 Spatial Mechanisms

As discussed in section 2.3.2, a large number of neurons in the primary visual cortex have receptive fields that resemble Gabor patterns (see Figure 2.10). Hence they can be characterized by a particular spatial frequency and orientation and essentially represent oriented band-pass filters. With a sufficient number of appropriately tuned cells, all orientations and frequencies in the sensitivity range of the visual system can be covered.

There is still a lot of discussion about the exact tuning shape and bandwidth, and different experiments have led to different results. For the achromatic visual pathways, most studies give estimates of 1–2 octaves for the spatial frequency bandwidth and 20–60 degrees for the orientation bandwidth, varying with spatial frequency (De Valois *et al.*, 1982a,b; Phillips and Wilson, 1984). These results are confirmed by psychophysical evidence from studies of discrimination and interaction phenomena (Olzak and Thomas, 1986). Interestingly, these cell properties can also be related with and even derived from the statistics of natural images (Field, 1987; van Hateren and van der Schaaf, 1998). Fewer empirical data are available for the chromatic pathways. They probably have similar spatial frequency band-widths (Webster *et al.*, 1990; Losada and Mullen, 1994, 1995), whereas their orientation bandwidths have been found to be significantly larger, ranging from 60 to 130 degrees (Vimal, 1997).

2.7.2 Temporal Mechanisms

Temporal mechanisms have been studied as well, but there is less agreement about their characteristics than for spatial mechanisms. While some studies concluded that there are a large number of narrowly tuned mechanisms (Lehky, 1985), it is now believed that there is just one low-pass and one band-pass mechanism (Watson, 1986; Hess and Snowden, 1992; Frederick-sen and Hess, 1998), which are generally referred to as *sustained* and *transient channel*, respectively. An additional third channel was proposed (Mandler and Makous, 1984; Hess and Snowden, 1992; Ascher and Gryz-wacz, 2000), but has been called in question by other studies (Hammett and Smith, 1992; Fredericksen and Hess, 1998). Fredericksen and Hess (1998) were able to achieve a very good fit to a large set of psychophysical data using one sustained and one transient mechanism. The frequency responses of the corresponding channels are shown in Figure 2.17.

Physiological experiments confirm these findings to the extent that low-pass and band-pass mechanisms have been discovered (Foster *et al.*, 1985),

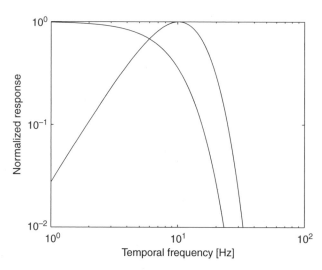

Figure 2.17 Temporal frequency responses of sustained (low-pass) and transient (band-pass) mechanisms of vision based on a model by Fredericksen and Hess (1997, 1998).

but neurons with band-pass properties exhibit a wide range of peak frequencies. Recent results also indicate that the peak frequency and bandwidth of the channels change considerably with stimulus energy (Fredericksen and Hess, 1997).

2.8 SUMMARY

Several important concepts of vision were presented. The major points can be summarized as follows:

- The human visual system is extremely complex. Our current knowledge is limited mainly to low-level processes.
- While the visual system is highly adaptive, it is not equally sensitive to all stimuli. There are a number of inherent limitations with respect to the visibility of stimuli.
- The response of the visual system depends much more on the contrast of patterns than on their absolute light levels.
- Visual information is processed in different pathways and channels in the visual system depending on its characteristics such as color, spatial and temporal frequency, orientation, phase, direction of motion, etc. These channels play an important role in explaining interactions between stimuli.
- Color perception is based on the different spectral sensitivities of photo-receptors and the decorrelation of their absorption rates into opponent colors.

These characteristics of the human visual system will be used in the design of vision models and quality metrics.

3

Video Quality

Beauty in things exists in the mind which contemplates them.

David Hume

The moving picture in all its incarnations (cinema, television, video, etc.) is one of the most widespread and most successful inventions of the twentieth century. In recent years, the development of powerful compression algorithms and video processing equipment has facilitated the move from the analog to the digital domain. Today, this move has already been completed in many stages of the video production and distribution chain. Reducing the bandwidth and storage requirements while maintaining a quality superior to that of analog video has been the priority in designing the new digital video systems, and guaranteeing a certain level of quality has become an important concern for content providers.

This chapter starts with an overview of video essentials, today's compression methods and standards. Compression and transmission of digital video entail a variety of characteristic artifacts and distortions, the most common of which are discussed here. Then we attempt to define and quantify visual quality from an observer's point of view and examine procedures for subjective quality assessment tests. Finally, we review the history and the state of the art of visual quality metrics, from simple pixel-based metrics such as MSE and PSNR to advanced vision-based metrics proposed in recent years.

Digital Video Quality - Vision Models and Metrics   Stefan Winkler
© 2005 John Wiley & Sons, Ltd   ISBN: 0-470-02404-6

3.1 VIDEO CODING AND COMPRESSION

Visual data in general and video in particular require large amounts of bandwidth and storage space. Uncompressed video at TV-resolution has typical data rates of a few hundred Mb/s, for example; for HDTV this goes up into the Gb/s range. Evidently, effective compression methods are vital to facilitate handling such data rates.

Compression is the reduction of redundancy in data. Generic lossless compression algorithms, which assure the perfect reconstruction of the initial data, could be used for images and video. However, these algorithms only achieve a data reduction of about 2:1 on average, which is not enough. When compressing video, two special types of redundancy can be exploited:

- *Spatio-temporal redundancy*: Typically, pixel values are correlated with their neighbors, both within the same frame and across frames.
- *Psychovisual redundancy*: The human visual system is not equally sensitive to all patterns (see Chapter 2). Therefore, the compression algorithm can discard information that is not visible to the observer. This is referred to as *lossy compression*.

In analog video, these two types of redundancies are exploited through vision-based color coding and interlacing techniques. Digital video offers additional compression methods, which are discussed afterwards.

3.1.1 Color Coding

Many compression schemes and video standards such as PAL, NTSC, or MPEG, are already based on human vision in the way that color information is processed. In particular, they take into account the nonlinear perception of lightness, the organization of color channels, and the low chromatic acuity of the human visual system (see Chapter 2).

Conventional television cathode ray tube (CRT) displays have a nonlinear, roughly exponential relationship between frame buffer RGB values or signal voltage and displayed intensity. In order to compensate for this, *gamma correction* is applied to the intensity values before coding. It so happens that the human visual system has an approximately logarithmic response to intensity, which is very nearly the inverse of the CRT nonlinearity (Poynton, 1998). Therefore, coding visual information in the gamma-corrected domain not only compensates for CRT behavior, but is also more meaningful perceptually.

The theory of opponent colors states that the human visual system decorrelates its input into white-black, red-green and blue-yellow difference signals, which are processed in separate visual channels (see section 2.5.2). Furthermore, chromatic visual acuity is significantly lower than achromatic acuity, as pointed out in section 2.4.2. In order to take advantage of this behavior, the color primaries red, green, and blue are rarely used for coding directly. Instead, *color difference* (chroma) signals similar to the ones just mentioned are computed. In component video, for example, the resulting color space is referred to as *YUV* or YC_BC_R, where *Y* encodes luminance, *U* or C_B the difference between the blue primary and luminance, and *V* or C_R the difference between the red primary and luminance.

The low chromatic acuity now permits a significant data reduction of the color difference signals. In digital video, this is achieved by *chroma subsampling*. The notation commonly used is as follows:

- 4:4:4 denotes no chroma subsampling.
- 4:2:2 denotes chroma subsampling by a factor of 2 horizontally; this sampling format is used in the standard for studio-quality component digital video as defined by ITU-R Rec. BT.601-5 (1995), for example.
- 4:2:0 denotes chroma subsampling by a factor of 2 both horizontally and vertically; it is probably the closest approximation of human visual color acuity achievable by chroma subsampling alone. This sampling format is the most common in JPEG or MPEG, e.g. for distribution-quality video.
- 4:1:1 denotes chroma subsampling by a factor of 4 horizontally.

3.1.2 Interlacing

As analog television was developed, it was noted that flicker could be perceived at certain frame rates, and that the magnitude of the flicker was a function of screen brightness and surrounding lighting conditions. A motion picture displayed in the theater at relatively low light levels can be displayed at a frame rate of 24 Hz. A bright CRT display requires a refresh rate of more than 50 Hz for flicker to disappear. The drawback of such a high frame rate is that the bandwidth of the signal becomes very large. On the other hand, the spatial resolution of the visual system decreases significantly at such temporal frequencies (this is the sharp fall-off range of the CSF in the high spatio-temporal frequency range, cf. Figure 2.13). These two properties combined gave rise to the technique referred to as *interlacing*.

The concept of interlacing is illustrated in Figure 3.1. Interlacing trades off vertical resolution against temporal resolution. Instead of sampling the video

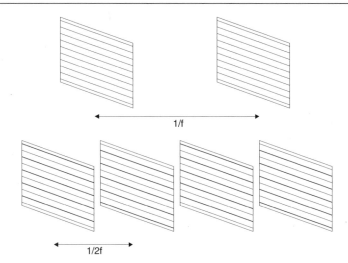

Figure 3.1 Illustration of interlacing. The top sequence is progressive: all lines of each frame are transmitted at the frame rate f. The bottom sequence is interlaced: each frame is split into two fields containing the odd and the even lines, respectively. These fields (bold lines) are transmitted alternately at twice the original frame rate (from S. Winkler *et al.* (2001), Vision and video: Models and applications, in C. J. van den Branden Lambrecht (ed.), *Vision Models and Applications to Image and Video Processing*, chap. 10, Kluwer Academic Publishers. Copyright © 2001 Springer. Used with permission.).

signal at 25 (PAL) or 30 (NTSC) frames per second, the sequence is shot at a frequency of 50 or 60 interleaved fields per second. A field corresponds to either the odd or the even lines of a frame, which are sampled at different time instants and displayed alternately. Thus the required bandwidth of the signal can be reduced by a factor of 2, while the full horizontal and vertical resolution is maintained for stationary image regions, and the refresh rate for objects larger than one scanline is still sufficiently high.

Interlacing is well suited to CRT display technology; LCD or plasma displays, however, are inherently progressive and require additional processing to handle interlaced material (de Haan and Bellers, 1998).

3.1.3 Compression Methods

As mentioned at the beginning of this section, digital video is amenable to special compression methods. They can be roughly classified into model-based methods, e.g. fractal compression, and waveform-based methods, e.g. DCT or wavelet compression. Most of today's video codecs and standards belong to the latter category and comprise the following stages (Tudor, 1995):

- *Transformation*: To facilitate exploiting psychovisual redundancies, the pictures are transformed to a domain where different frequency ranges with varying sensitivities of the human visual system can be separated. This can be achieved by the discrete cosine transform (DCT) or the wavelet transform, for example. This step is reversible, i.e. no information is lost.

- *Quantization*: After the transformation, the numerical precision of the transform coefficients is reduced in order to decrease the number of bits in the stream. The degree of quantization applied to each coefficient is usually determined by the visibility of the resulting distortion to a human observer; high-frequency coefficients can be more coarsely quantized than low-frequency coefficients, for example. Quantization is the stage that is responsible for the 'lossy' part of compression.

- *Coding*: After the data has been quantized into a finite set of values, it can be encoded losslessly by exploiting the redundancy between the quantized coefficients in the bitstream. Entropy coding, which relies on the fact that certain symbols occur much more frequently than others, is often used for this process. Two of the most popular entropy coding schemes are Huffman coding and arithmetic coding (Sayood, 2000).

A key aspect of digital video compression is exploiting the similarity between successive frames in a sequence instead of coding each picture separately. While this temporal redundancy could be taken care of by a spatio-temporal transformation, a hybrid spatial- and transform-domain approach is often adopted instead for reasons of implementation efficiency. A simple method for temporal compression is frame differencing, where only the pixel-wise differences between successive frames are coded. Higher compression can be achieved using *motion estimation*, a technique for describing a frame based on the content of nearby frames with the help of motion vectors. By compensating for the movements of objects in this manner, the differences between frames can be further reduced.

3.1.4 Standards

The Moving Picture Experts Group (MPEG)[†] is a working group of ISO/IEC in charge of developing international standards for the compression, decompression, processing, and coded representation of moving pictures, audio, and their combination. MPEG comprises some of the most popular and

[†]See http://www.chiariglione.org/mpeg/ for an overview of its activities.

widespread standards for video coding. The group was established in January 1988, and since then it has produced:

- MPEG-1, a standard for storage and retrieval of moving pictures and audio, which was approved in 1992. MPEG-1 defines a block-based hybrid DCT/DPCM coding scheme with prediction and motion compensation. It also provides functionality for random access in digital storage media.
- MPEG-2, a standard for digital television, which was approved in 1994. The video coding scheme of MPEG-2 is a refinement of MPEG-1. Special consideration is given to interlaced sources. Furthermore, many function-alities such as scalability were introduced. In order to keep implementa-tion complexity low for products not requiring all video formats supported by the standard, so-called 'Profiles', describing functionalities, and 'Levels', describing parameter constraints such as resolutions and bitrates, were defined to provide separate MPEG-2 conformance levels.
- MPEG-4, a standard for multimedia applications, whose parts one and two (video and systems) were approved in 1998. MPEG-4 addresses the need for robustness in error-prone environments, interactive functionality for content-based access and manipulation, and a high compression efficiency at very low bitrates. MPEG-4 achieves these goals by means of an object-oriented coding scheme using so-called 'audio-visual objects', for exam-ple a fixed background, the picture of a person in front of that background, the voice associated with that person etc.
- MPEG-4 part 10, Advanced Video Coding (AVC), also known as ITU-T Rec. H.264 (2003).[†] This latest standard is designed for a wide range of applications, ranging from from mobile video to HDTV. It is based on the same general block-based hybrid coding approach as the other MPEG standards. The new features include smaller block sizes, more flexible prediction both temporally (inter-frame) and spatially (intra-frame), an in-loop deblocking filter to reduce the visibility of the characteristic blocking artifacts, and further improved error resilience. All these incremental improvements together result in an approximately two times higher coding efficiency compared to previous standards.

The two other standards in this family, MPEG-7 and MPEG-21, are not about codecs and are thus of less interest here. MPEG-7 is a standard for content description in the context of audio-visual information indexing, search and retrieval, and was approved in 2001. MPEG-21 is concerned

---

[†]In older documents it is sometimes referred to as H.26L or JVT codec.

with interoperability between the elements of a multimedia application infrastructure (mainly devices and content) and defines how they should relate, integrate, and interact; its different parts will be standardized from 2004 onwards.

MPEG coding standards are intended to be generic, i.e. only the bitstream syntax is defined, and therefore mainly the decoding scheme is standardized. The design of the encoder is left up to the implementor.

MPEG-2 is one of the most widespread standards in commercial use today. It is used on DVDs as well as for digital TV and HDTV broadcast. We will therefore look at MPEG-2 video compression a bit more closely. The essentials are quite similar for the other MPEG video standards.

An MPEG-2 video stream is hierarchically structured, as illustrated in Figure 3.2 (Tudor, 1995). The sequence is composed of three types of frames,

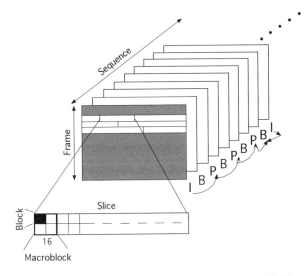

Figure 3.2 Elements of an MPEG-2 video sequence (from S. Winkler *et al.* (2001), Vision and video: Models and applications, in C. J. van den Branden Lambrecht (ed.), *Vision Models and Applications to Image and Video Processing*, chap. 10, Kluwer Academic Publishers. Copyright © 2001 Springer. Used with permission.).

namely intra-coded (I), forward predicted (P), and bidirectionally predicted (B) frames. Each frame is subdivided into slices, which are a collection of consecutive macroblocks. Each macroblock in turn contains four blocks of 8×8 pixels each. The DCT is computed on these blocks, while motion estimation is performed on macroblocks. The resulting DCT coefficients are quantized and variable-length coded.

The MPEG-2 system specification defines a multiplexed structure for combining audio and video data as well as timing information for transmission over a communication channel. It is based on two levels of packetization. First, the compressed bitstreams or *elementary streams* (audio or video) are packetized. Subsequently, the packetized elementary streams are multiplexed together to create the *transport stream*, which can carry multiple audio and video programs.[†] It consists of fixed-size packets of 188 bytes each; their headers contain synchronization and timing information. Finally, the transport stream is encapsulated in real-time protocol (RTP) packets for transmission.

Other standards being used commercially today are MPEG-1 (on VCDs) and ITU-T Rec. H.263 (1998) (for video conferencing). Third-generation (3G) mobile video phones will rely mainly on MPEG-4 and H.263 codecs. Digital video camcorders use DV, an intra-frame block-DCT based coding scheme (similar to Motion-JPEG); it is an IEC and SMPTE standard.

The recent surge of multimedia applications has led to the development of a large variety of additional compression/decompression methods; Real Media Video[‡] and Windows Media Video[§] are among the best-known. These codecs are based on the discrete cosine transform, the wavelet transform, vector quantization, or combinations thereof. In contrast to MPEG, however, most of them are proprietary.

For a more detailed overview of video compression technologies the reader is referred to Symes (2003).

3.2 ARTIFACTS

3.2.1 Compression Artifacts

As pointed out in section 3.1.4, the compression algorithms used in various video coding standards are quite similar. Most of them rely on motion compensation and block-based DCT with subsequent quantization of the coefficients. In such coding schemes, compression distortions are caused by only one operation, namely the quantization of the transform coefficients. Although other factors affect the visual quality of the stream, such as motion prediction or decoding buffer size, they do not introduce any distortion per se, but affect the encoding process indirectly.

[†]In error-free environments, a program stream (without additional packetization) may be used instead.
[‡]http://www.realnetworks.com/products/codecs/realvideo.html
[§]http://www.microsoft.com/windows/windowsmedia/9series/codecs/video.aspx

A variety of artifacts can be distinguished in a compressed video sequence (Yuen and Wu, 1998):

- The *blocking effect* or blockiness refers to a block pattern in the compressed sequence. It is due to the independent quantization of individual blocks (usually of 8×8 pixels in size) in block-based DCT coding schemes, leading to discontinuities at the boundaries of adjacent blocks. The blocking effect is often the most prominent visual distortion in a compressed sequence due to the regularity and extent of the pattern (see Figure 3.3(b)). Recent codecs such as H.264 employ a deblocking filter to reduce the visibility of this artifact.

      (a) Original   (b) Block-DCT   (c) Wavelet

Figure 3.3 Illustration of typical compression artifacts for block-DCT based methods (b) and wavelet-based methods (c). The blocking effect and DCT basis images are clearly visible in the bottom part of (b); the staircase effect can be seen around the white slanted edge of the lighthouse in (b). Blur is evident in both compressed images; ringing can be observed around contours and edges.

- *Blur* manifests itself as a loss of spatial detail and a reduction of edge sharpness. It is due to the suppression of the high-frequency coefficients by coarse quantization (see Figure 3.3).
- *Color bleeding* is the smearing of colors between areas of strongly differing chrominance. It results from the suppression of high-frequency coefficients of the chroma components. Due to chroma subsampling, color bleeding extends over an entire macroblock.
- The *DCT basis image effect* is prominent when a single DCT coefficient is dominant in a block. At coarse quantization levels, this results in an emphasis of the dominant basis image and the reduction of all other basis images (see Figure 3.3(b)).
- Slanted lines often exhibit the *staircase effect.* It is due to the fact that DCT basis images are best suited to the representation of horizontal and vertical lines, whereas lines with other orientations require higher-frequency DCT coefficients for accurate reconstruction. The typically strong quantization of these coefficients causes slanted lines to appear jagged (see Figure 3.3(b)).
- *Ringing* is fundamentally associated with Gibbs' phenomenon and is thus most evident along high-contrast edges in otherwise smooth areas. It is a direct result of quantization leading to high-frequency irregularities in the reconstruction. Ringing occurs with both luminance and chroma components (see Figure 3.3).
- *False edges* are a consequence of the transfer of block-boundary discontinuities (due to the blocking effect) from reference frames into the predicted frame by motion compensation.
- *Jagged motion* can be due to poor performance of the motion estimation. Block-based motion estimation works best when the movement of all pixels in a macroblock is identical. When the residual error of motion prediction is large, it is coarsely quantized.
- Motion estimation is often conducted with the luminance component only, yet the same motion vector is used for the chroma components. This can result in *chrominance mismatch* for a macroblock.
- *Mosquito noise* is a temporal artifact seen mainly in smoothly textured regions as luminance/chrominance fluctuations around high-contrast edges or moving objects. It is a consequence of the coding differences for the same area of a scene in consecutive frames of a sequence.
- *Flickering* appears when a scene has high texture content. Texture blocks are compressed with varying quantization factors over time, which results in a visible flickering effect.
- *Aliasing* can be noticed when the content of the scene is above the Nyquist rate, either spatially or temporally.

While some of these effects are unique to block-based coding schemes, many of them are observed with other compression algorithms as well. In wavelet-based compression, for example, the transform is applied to the entire image, therefore none of the block-related artifacts occur. Instead, blur and ringing are the most prominent distortions (see Figure 3.3(c)).

3.2.2 Transmission Errors

An important and often overlooked source of impairments is the transmission of the bitstream over a noisy channel. Digitally compressed video is typically transferred over a packet-switched network. The physical transport can take place over a wire or wireless, where some transport protocol such as ATM or TCP/IP ensures the transport of the bitstream. The bitstream is transported in packets whose headers contain sequencing and timing information. This process is illustrated in Figure 3.4. Streams can carry additional signaling

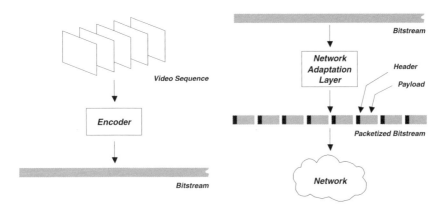

Figure 3.4 Illustration of a video transmission system. The video sequence is first compressed by the encoder. The resulting bitstream is packetized in the network adaptation layer, where a header containing sequencing and synchronization data is added to each packet. The packets are then sent over the network (from S. Winkler *et al.* (2001), Vision and video: Models and applications, in C. J. van den Branden Lambrecht (ed.), *Vision Models and Applications to Image and Video Processing*, chap. 10, Kluwer Academic Publishers. Copyright © 2001 Springer. Used with permission.).

information at the session level. A variety of protocols are used to transport the audio-visual information, synchronize the actual media and add timing information. Most applications require the streaming of video, i.e. it must be possible to decode and display the bitstream in real time as it arrives.

Two different types of impairments can occur when transporting media over noisy channels. Packets may be corrupted and thus discarded, or they may be delayed to the point where they are not received in time for decoding. The latter is due to the packet routing and queuing algorithms in routers and switches. To the application, both have the same effect: part of the media stream is not available, thus packets are missing when they are needed for decoding.

Such losses can affect both the semantics and the syntax of the media stream. When the losses affect syntactic information, not only the data relevant to the lost block are corrupted, but also any other data that depend on this syntactic information. For example, an MPEG macroblock that is damaged through the loss of packets corrupts all following macroblocks until an end of slice is encountered, where the decoder can resynchronize. This spatial loss propagation is due to the fact that the DC coefficient of a macroblock is differentially predicted between macroblocks and reset at the beginning of a slice. Furthermore, for each of these corrupted macroblocks, all blocks that are predicted from them by motion estimation will be damaged as well, which is referred to as temporal loss propagation. Hence the loss of a single macroblock can affect the stream up to the next intra-coded frame. These loss propagation phenomena are illustrated in Figure 3.5. H.264 introduces flexible macroblock ordering to alleviate this problem: the

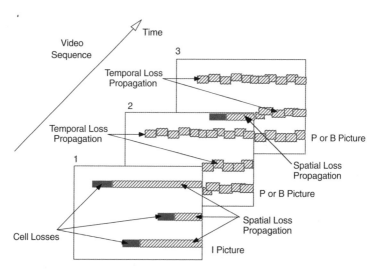

Figure 3.5 Spatial and temporal propagation of losses in an MPEG-compressed video sequence. The loss of a single macroblock causes the inability to decode the data up to the end of the slice. Macroblocks in neighboring frames that are predicted from the damaged area are corrupted as well (from S. Winkler *et al.* (2001), Vision and video: Models and applications, in C. J. van den Branden Lambrecht (ed.), *Vision Models and Applications to Image and Video Processing*, chap. 10, Kluwer Academic Publishers. Copyright © 2001 Springer. Used with permission.).

encoded bits describing neighboring macroblocks in the video can be put in different parts of the bitstream, thus spreading the errors more evenly across the frame or video.

The effect can be even more damaging when global data are corrupted. An example of this is the timing information in an MPEG stream. The system layer specification of MPEG imposes that the decoder clock be synchronized with the encoder clock via periodic refresh of the program clock reference sent in some packet. Too much jitter on packet arrival can corrupt the synchronization of the decoder clock, which can result in highly noticeable impairments.

The visual effects of such losses vary significantly between decoders depending on their ability to deal with corrupted streams. Some decoders never recover from certain errors, while others apply concealment techniques such as early synchronization or spatial and temporal interpolation in order to minimize these effects (Wang and Zhu, 1998).

3.2.3 Other Impairments

Aside from compression artifacts and transmission errors, the quality of digital video sequences can be affected by any pre- or post-processing stage in the system. These include:

- conversions between the digital and the analog domain;
- chroma subsampling (discussed in section 3.1.1);
- frame rate conversion between different display formats;
- de-interlacing, i.e. the process of creating a progressive sequence from an interlaced one (de Haan and Bellers, 1998; Thomas, 1998).

One particular example is the so-called 3:2 pulldown, which denotes the standard way to convert progressive film sequences shot at 24 frames per second to interlaced video at 60 fields per second.

3.3 VISUAL QUALITY

3.3.1 Viewing Distance

For studying visual quality, it is helpful to relate system and setup parameters to the human visual system. For instance, it is very popular in the video community to specify viewing distance in terms of display size, i.e. in multiples of screen height. There are two reasons for this: first, it was assumed for quite some time that the ratio of preferred viewing distance to screen height is constant (Lund, 1993). However, more recent experiments with larger displays have shown that this is not the case. While the preferred viewing distance is indeed around 6–7 screen heights or more for smaller displays, it approaches 3–4 screen heights with increasing display size (Ardito *et al.*, 1996; Lund, 1993). Incidentally, typical home viewing distances are far from ideal in this respect (Alpert, 1996). The second reason was the implicit assumption of a certain display resolution (a certain number of scan lines), which is usually fixed for a given television standard.

In the context of vision modeling, the size and resolution of the image projected onto the retina are more adequate specifications (see section 2.1.1). For a given screen height H and viewing distance D, the size is measured in degrees of visual angle α:

$$\alpha = 2 \operatorname{atan} (H/2D). \tag{3.1}$$

The resolution or maximum spatial frequency f_{max} is measured in cycles per degree of visual angle (cpd). It is computed from the number of scan lines L according to the Nyquist sampling theorem:

$$f_{max} = L/2\alpha \, [\text{cpd}]. \tag{3.2}$$

The size and resolution of the image that popular video formats produce on the retina are shown in Figure 3.6 for a typical range of viewing distances and screen heights. It is instructive to compare them to the corresponding 'specifications' of the human visual system mentioned in Chapter 2.

For example, from the contrast sensitivity functions shown in Figure 2.13 it is evident that the scan lines of PAL and NTSC systems at viewing distances below 3–4 screen heights ($f_{max} \approx 15 \, \text{cpd}$) can easily be resolved by the viewer. HDTV provides approximately twice the resolution and is thus better suited for close viewing and large screens.

3.3.2 Subjective Quality Factors

In order to be able to design reliable visual quality metrics, it is necessary to understand what 'quality' means to the viewer (Ahumada and Null, 1993; Klein, 1993; Savakis *et al.*, 2000). Viewers' enjoyment when watching a video depends on many factors:

- *Individual interests and expectations*: Everyone has their favorite programs, which implies that a football fan who attentively follows a game may have very different quality requirements than someone who is only marginally interested in the sport. We have also come to expect different

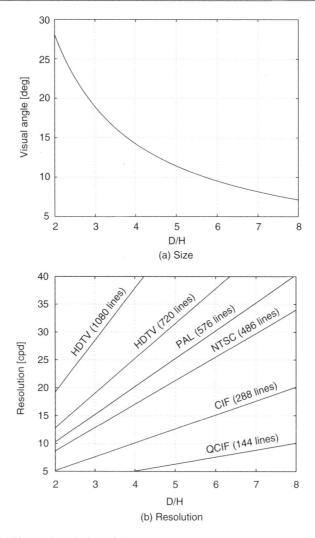

Figure 3.6 Size and resolution of the image that popular video formats produce on the retina as a function of viewing distance D in multiples of screen height H.

qualities in different situations, e.g. the quality of watching a feature film at the cinema versus a short clip on a mobile phone. At the same time, advances in technology such as the DVD have raised the quality bar – a VHS recording that nobody would have objected to a few years ago is now considered inferior quality by everyone who has a DVD player at home.

- *Display type and properties*: There is a wide variety of displays available today – traditional CRT screens, LCDs, plasma displays, front and back projection technologies. They have different characteristics in terms of brightness, contrast, color rendition, response time etc., which determine the quality of video rendition. Compression artifacts (especially blockiness) are more visible on non-CRT displays, for example (EBU BTMC, 2002; Pinson and Wolf, 2004). As already discussed in section 3.3.1, display resolution and size (together with the viewing distance) also influence perceived quality (Westerink and Roufs, 1989; Lund, 1993).

- *Viewing conditions*: Aside from the viewing distance, the ambient light affects our perception to a great extent. Even though we are able to adapt to a wide range of light levels and to discount the color of the illumination, high ambient light levels decrease our sensitivity to small contrast variations. Furthermore, exterior light can lead to veiling glare due to reflections on the screen that again reduce the visible luminance and contrast range (Süsstrunk and Winkler, 2004).

- *The fidelity of the reproduction*. On the one hand, we want the 'original' video to arrive at the end-user with a minimum of distortions introduced along the way. On the other hand, video is not necessarily about capturing and reproducing a scene as naturally as possible – think of animations, special effects or artistic 'enhancements'. For example, sharp images with high contrast are usually more appealing to the average viewer (Roufs, 1989). Likewise, subjects prefer slightly more colorful and saturated images despite realizing that they look somewhat unnatural (de Ridder *et al.*, 1995; Fedorovskaya *et al.*, 1997; Yendrikhovskij *et al.*, 1998). These phenomena are well understood and utilized by professional photographers (Andrei, 1998, personal communication; Marchand, 1999, personal communication).

- Finally, the accompanying *soundtrack* has a great influence on perceived quality of the viewing experience (Beerends and de Caluwe, 1999; Joly *et al.*, 2001; Winkler and Faller, 2005). Subjective quality ratings are generally higher when the test scenes are accompanied by good quality sound (Rihs, 1996). Furthermore, it is important that the sound be synchronized with the video. This is most noticeable for speech and lip synchronization, for which time lags of more than approximately 100 ms are considered very annoying (Steinmetz, 1996).

Unfortunately, subjective quality cannot be represented by an exact figure; due to its inherent subjectivity, it can only be described statistically. Even in psychophysical threshold experiments, where the task of the observer is just to give a yes/no answer, there exists a significant variation in contrast sensitivity functions and other critical low-level visual parameters between different observers. When the artifacts become supra-threshold, the observers are bound to apply different weightings to each of them. Deffner *et al.* (1994) showed that experts and non-experts (with respect to image quality) examine different critical image characteristics to form their opinion. With all these caveats in mind, testing procedures for subjective quality assessment are discussed next.

3.3.3 Testing Procedures

Subjective experiments represent the benchmark for vision models in general and quality metrics in particular. However, different applications require different testing procedures. Psychophysics provides the tools for measuring the perceptual performance of subjects (Gescheider, 1997; Engeldrum, 2000).

Two kinds of decision tasks can be distinguished, namely *adjustment* and *judgment* (Pelli and Farell, 1995). In the former, the observer is given a classification and provides a stimulus, while in the latter, the observer is given a stimulus and provides a classification. Adjustment tasks include setting the threshold amplitude of a stimulus, cancelling a distortion, or matching a stimulus to a given one. Judgment tasks on the other hand include yes/no decisions, forced choices between two alternatives, and magnitude estimation on a rating scale.

It is evident from this list of adjustment and judgment tasks that most of them focus on threshold measurements. Traditionally, the concept of threshold has played an important role in psychophysics. This has been motivated by the desire to minimize the influence of perception and cognition by using simple criteria and tasks. Signal detection theory has provided the statistical framework for such measurements (Green and Swets, 1966). While such threshold detection experiments are well suited to the investigation of low-level sensory mechanisms, a simple yes/no answer is not sufficient to capture the observer's experience in many cases, including visual quality assessment. This has stimulated a great deal of experimentation with supra-threshold stimuli and non-detection tasks.

Subjective testing for visual quality assessment has been formalized in ITU-R Rec. BT.500-11 (2002) and ITU-T Rec. P.910 (1999), which suggest standard viewing conditions, criteria for the selection of observers and test material, assessment procedures, and data analysis methods. ITU-R Rec. BT.500-11 (2002) has a longer history and was written with television applications in mind, whereas ITU-T Rec. P.910 (1999) is intended for multimedia applications. Naturally, the experimental setup and viewing conditions differ in the two recommendations, but the procedures from both should be considered for any experiment.

The three most commonly used procedures from ITU-R Rec. BT.500-11 (2002) are the following:

- Double Stimulus Continuous Quality Scale (DSCQS). The presentation sequence for a DSCQS trial is illustrated in Figure 3.7(a). Viewers are

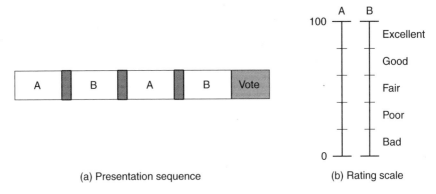

(a) Presentation sequence   (b) Rating scale

Figure 3.7 DSCQS method. The reference and the test sequence are presented twice in alternating fashion (a). The order of the two is chosen randomly for each trial, and subjects are not informed which is which. They rate each of the two separately on a continuous quality scale ranging from 'bad' to 'excellent' (b).

shown multiple sequence pairs consisting of a 'reference' and a 'test' sequence, which are rather short (typically 10 seconds). The reference and test sequence are presented twice in alternating fashion, with the order of the two chosen randomly for each trial. Subjects are not informed which is the reference and which is the test sequence. They rate each of the two separately on a continuous quality scale ranging from 'bad' to 'excellent' as shown in Figure 3.7(b). Analysis is based on the difference in rating for each pair, which is calculated from an equivalent numerical scale from 0 to 100. This differencing helps reduce the subjectivity with respect to scene content and experience. DSCQS is the preferred method when the quality of test and reference sequence are similar, because it is quite sensitive to small differences in quality.

- Double Stimulus Impairment Scale (DSIS). The presentation sequence for a DSIS trial is illustrated in Figure 3.8(a). As opposed to the DSCQS method, the reference is always shown before the test sequence, and

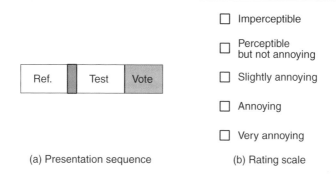

(a) Presentation sequence   (b) Rating scale

Figure 3.8 DSIS method. The reference and the test sequence are shown only once (a). Subjects rate the amount of impairment in the test sequence on a discrete five-level scale ranging from 'very annoying' to 'imperceptible' (b)

neither is repeated. Subjects rate the amount of impairment in the test sequence on a discrete five-level scale ranging from 'very annoying' to 'imperceptible' as shown in Figure 3.8(b). The DSIS method is well suited for evaluating clearly visible impairments such as artifacts caused by transmission errors.

- Single Stimulus Continuous Quality Evaluation (SSCQE) (MOSAIC, 1996). Instead of seeing separate short sequence pairs, viewers watch a program of typically 20–30 minutes' duration which has been processed by the system under test; the reference is not shown. Using a slider, the subjects continuously rate the instantaneously perceived quality on the DSCQS scale from 'bad' to 'excellent'.

ITU-T Rec. P.910 (1999) defines the following testing procedures:

- Absolute Category Rating (ACR). This is a single stimulus method; viewers only see the video under test, without the reference. They give one rating for its overall quality using a discrete five-level scale from 'bad' to 'excellent'. The fact that the reference is not shown with every test clip makes ACR a very efficient method compared to DSIS or DSCQS, which take almost 2 or 4 times as long, respectively.
- Degradation Category Rating (DCR), which is identical to DSIS.
- Pair Comparison (PC). For this method, test clips from the same scene but different conditions are paired in all possible combinations, and viewers make a preference judgment for each pair. This allows very fine quality discrimination between clips.

For all of these methods, the ratings from all observers (a minimum of 15 is recommended) are then averaged into a Mean Opinion Score (MOS),[†] which represents the subjective quality of a given clip.

The testing procedures mentioned above generally have different applications. All single-rating methods (DSCQS, DSIS, ACR, DCR, PC) share a common drawback, however: changes in scene complexity, statistical multiplexing or transmission errors can produce substantial quality variations that are not evenly distributed over time; severe degradations may appear only once every few minutes. Single-rating methods are not suited to the evaluation of such long sequences because of the recency effect, a bias in the ratings toward the final 10–20 seconds due to limitations of human working memory (Aldridge *et al.*, 1995). Furthermore, it has been argued that the presentation of a reference or the repetition of the sequences in the DSCQS method puts the subjects in a situation too removed from the home viewing environment by allowing them to become familiar with the material under investigation (Lodge, 1996). SSCQE has been designed with these problems in mind, as it relates well to the time-varying quality of today's compressed digital video systems (MOSAIC, 1996). On the other hand, program content tends to have an influence on SSCQE scores. Also, SSCQE ratings are more difficult to handle in the analysis because of the potential differences in viewer reaction times and the inherent autocorrelation of time-series data.

3.4 QUALITY METRICS

3.4.1 Pixel-based Metrics

The mean squared error (MSE) and the peak signal-to-noise ratio (PSNR) are the most popular difference metrics in image and video processing. The MSE is the mean of the squared differences between the gray-level values of pixels in two pictures or sequences I and $\tilde{I}$:

$$\text{MSE} = \frac{1}{TXY} \sum_t \sum_x \sum_y [I(t,x,y) - \tilde{I}(t,x,y)]^2 \tag{3.3}$$

for pictures of size $X \times Y$ and T frames in the sequence. The root mean squared error is simply $\text{RMSE} = \sqrt{\text{MSE}}$.

[†]Differential Mean Opinion Score (DMOS) in the case of DSCQS.

The PSNR in decibels is defined as:

$$\text{PSNR} = 10 \log \frac{m^2}{\text{MSE}}, \tag{3.4}$$

where m is the maximum value that a pixel can take (e.g. 255 for 8-bit images). Note that MSE and PSNR are well defined only for luminance information; once color comes into play, there is no agreement on the computation of these measures.

Technically, MSE measures image difference, whereas PSNR measures image fidelity, i.e. how closely an image resembles a reference image, usually the uncorrupted original. The popularity of these two metrics is rooted in the fact that minimizing the MSE is equivalent to least-squares optimization in a minimum energy sense, for which well-known mathematical tools are readily available. Besides, computing MSE and PSNR is very easy and fast. Because they are based on a pixel-by-pixel comparison of images, however, they only have a limited, approximate relationship with the distortion or quality perceived by the human visual system. In certain situations the subjective image quality can be improved by adding noise and thereby reducing the PSNR. Dithering of color images with reduced color depth, which adds noise to the image to remove the perceived banding caused by the color quantization, is a common example of this. Furthermore, the visibility of distortions depends to a great extent on the image background, a property known as masking (see section 2.6.1). Distortions are often much more disturbing in relatively smooth areas of an image than in texture regions with a lot of activity, an effect not taken into account by pixel-based metrics. Therefore the perceived quality of images with the same PSNR can actually be very different. An example of the problems with using PSNR as a quality indicator is shown in Figure 3.9.

A number of additional pixel-based metrics are discussed by Eskicioglu and Fisher (1995). They found that although some of these metrics can predict subjective ratings quite successfully for a given compression technique or type of distortion, they are not reliable for evaluations across techniques. Another study by Marmolin (1986) concluded that even perceptual weighting of MSE does not give consistently reliable predictions of visual quality for different pictures and scenes. These results indicate that pixel-based error measures are not accurate for quality evaluations across different scenes or distortion types. Therefore it is imperative for reliable quality metrics to consider the way the human visual system processes visual information.

(a) Original   (b) PSNR = 32 dB   (c) PSNR = 32 dB

Figure 3.9 The same amount of noise was inserted into images (b) and (c) such that their PSNR with respect to the original (a) is identical. Band-pass filtered noise was inserted into the top region of image (b), whereas high-frequency noise was inserted into the bottom region of image (c). Our sensitivity to the structured (low-frequency) noise in image (b) is already quite high, and it is clearly visible on the smooth sky background. The noise in image (c) is hardly detectable due to our low sensitivity for high-frequency stimuli and the strong masking by highly textured content in the bottom region. PSNR is oblivious to both of these effects.

In the following, the implementation and performance of a variety of quality metrics are discussed. Because of the abundance of quality metrics described in the literature, only a limited number have been selected for this review. In particular, we focus on single- and multi-channel models of vision. A generic block diagram that applies to most of the metrics discussed here is shown in Figure 3.10 (of course, not all blocks are implemented by all metrics). The characteristics of these and a few other quality metrics are summarized at the end of the section in Table 3.1. The modeling details of the different metric components will be discussed later in Chapter 4.

3.4.2 Single-channel Models

The first models of human vision adopted a single-channel approach. Single-channel models regard the human visual system as a single spatial filter,

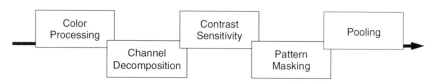

Figure 3.10 Generic block diagram of a vision-based quality metric. The input image or video typically undergoes color processing, which may include color space conversion and lightness transformations, a decomposition into a number of visual channels (for multi-channel models), application of the contrast sensitivity function, a model of pattern masking, and pooling of the data from the different channels and locations.

whose characteristics are defined by the contrast sensitivity function. The output of such a system is the filtered version of the input stimulus, and detectability depends on a threshold criterion.

The first computational model of vision was designed by Schade (1956) to predict pattern sensitivity for foveal vision. It is based on the assumption that the cortical representation is a shift-invariant transformation of the retinal image and can thus be expressed as a convolution. In order to determine the convolution kernel of this transformation, Schade carried out psychophysical experiments to measure the sensitivity to harmonic contrast patterns. From this CSF, the convolution kernel for the model can be computed, which is an estimate of the psychophysical line spread function (see section 2.1.3). Schade's model was able to predict the visibility of simple stimuli but failed as the complexity of the patterns increased.

The first image quality metric for luminance images was developed by Mannos and Sakrison (1974). They realized that simple pixel-based distortion measures were not able to accurately predict the quality differences perceived by observers. On the basis of psychophysical experiments on the visibility of gratings, they inferred some properties of the human visual system and came up with a closed-form expression for contrast sensitivity as a function of spatial frequency, which is still widely used in HVS-models. The input images are filtered with this CSF after a lightness nonlinearity. The squared difference between the filter output for the two images is the distortion measure. It was shown to correlate quite well with subjective ranking data. Albeit simple, this metric was one of the first works in engineering to recognize the importance of applying vision science to image processing.

The first color image quality metric was proposed by Faugeras (1979). His model computes the cone absorption rates and applies a logarithmic nonlinearity to obtain the cone responses. One achromatic and two chromatic color difference components are calculated from linear combinations of the cone responses to account for the opponent-color processes in the human visual system. These opponent-color signals go through individual filtering stages with the corresponding CSFs. The squared differences between the resulting filtered components for the reference image and the distorted image are the basis for an estimate of image distortion.

The first video quality metric was developed by Lukas and Budrikis (1982). It is based on a spatio-temporal model of the contrast sensitivity function using an excitatory and an inhibitory path. The two paths are combined in a nonlinear way, enabling the model to adapt to changes in the level of background luminance. Masking is also incorporated in the model by means of a weighting function derived from the spatial and temporal activity in the reference sequence. In the final stage of the metric, an L_p-norm of the masked error signal is computed over blocks in the frame whose size is chosen such that each block covers the size of the foveal field of vision. The resulting distortion measure was shown to outperform MSE as a predictor of perceived quality.

Tong $et\ al.$ (1999) proposed an interesting single-channel video quality metric called ST-CIELAB (spatio-temporal CIELAB). ST-CIELAB is an extension of the spatial CIELAB (S-CIELAB) image quality metric (Zhang and Wandell, 1996). Both are backward compatible to the CIELAB standard, i.e. they reduce to CIE $L^*a^*b^*$ (see Appendix) for uniform color fields. The ST-CIELAB metric is based on a spatial, temporal, and chromatic model of human contrast sensitivity in an opponent color space. The outputs of this model are transformed to CIE $L^*a^*b^*$ space, whose ΔE difference formula (equation (A.6)) is then used for pooling.

Single-channel models and metrics are still in use because of their relative simplicity and computational efficiency, and a variety of extensions and improvements have been proposed. However, they are intrinsically limited in prediction accuracy. They are unable to cope with more complex patterns and cannot account for empirical data from masking and pattern adaptation experiments (see section 2.6). These data can be explained quite successfully by a multi-channel theory of vision, which assumes a whole set of different channels instead of just one. The corresponding multi-channel models and metrics are discussed in the next section.

3.4.3 Multi-channel Models

Multi-channel models assume that each band of spatial frequencies is dealt with by a separate channel (see section 2.7). The CSF is essentially the envelope of the sensitivities of these channels. Detection occurs independently in any channel when the signal in that band reaches a threshold.

Watson (1987a) introduced the cortex transform, a multi-resolution pyramid that simulates the spatial-frequency and orientation tuning of simple cells in the primary visual cortex (see section 2.3.2). It is appealing because of its flexibility: spatial frequency selectivity and orientation selectivity are modeled separately, the filter bandwidths can be adjusted within a broad range, and the transform is easily invertible. Watson and Ahumada (1989) later proposed an orthogonal-oriented pyramid operating on a hexagonal lattice as an alternative decomposition tool.

Watson (1987b) used the cortex transform in a spatial model for luminance image coding, where it serves as the first analysis and decomposition stage. Pattern sensitivity is then modeled with a contrast sensitivity function and intra-channel masking. A perceptual quantizer is used to compress the filtered signals for minimum perceptual error.

Watson (1990) was also the first to outline the architecture of a multi-channel vision model for video coding. It is a straightforward extension of the above-mentioned spatial model for still images (Watson, 1987b). The model partitions the input into achromatic and chromatic opponent-color channels, into static and motion channels, and further into channels of particular frequencies and orientations. Bits are then allocated to each band taking into account human visual sensitivity to that band as well as visual masking effects. In contrast to the spatial model for images, it has never been implemented and tested, however.

Daly (1993) proposed the Visual Differences Predictor (VDP), a rather well-known image distortion metric. The underlying vision model includes an amplitude nonlinearity to account for the adaptation of the visual system to different light levels, an orientation-dependent two-dimensional CSF, and a hierarchy of detection mechanisms. These mechanisms involve a decomposition similar to the above-mentioned cortex transform and a simple intra-channel masking function. The responses in the different channels are converted to detection probabilities by means of a psychometric function and finally combined according to rules of probability summation. The resulting output of the VDP is a visibility map indicating the areas where two images differ in a perceptual sense.

Lubin (1995) designed the Sarnoff Visual Discrimination Model (VDM) for measuring still image fidelity. First the input images are convolved with an approximation of the point spread function of the eye's optics. Then the sampling by the cone mosaic on the retina is simulated. The decomposition stage implements a Laplacian pyramid for spatial frequency separation, local

Table 3.1 Overview of visual quality metrics

Reference	Appl.[1]	Color Space[2]	Lightness[3]	Transform[4]	Local Contrast	CSF[5]	Masking[6]	Pooling[7]	Eval.[8]	Comments
Mannos and Sakrison (1974)	IQ, IC	Lum.	$L^{0.33}$			F		L_2	R	
Faugeras (1979)	IQ, IC	AC_1C_2	log L			F		L_2	E	
Lukas and Budrikis (1982)	VQ	Lum.			yes	F	C	L_p	R	
Girod (1989)	VQ	Lum.			yes	F	C	L_2, L_∞	R	Integral spatio-temporal model
Malo et al. (1997)	IQ	Lum.	?			F		L_2	R	DCT-based error weighting
Zhang and Wandell (1996)	IQ	Opp.	$L^{1/3}$	Fourier		F			E	Spatial CIELAB extension
Tong et al. (1999)	VQ	Opp.	$L^{1/3}$	Fourier		F		L_1	R	Spatio-temporal CIELAB extension
Daly (1993)	IQ	Lum.	yes	mod. Cortex		F	C	PS	E	Visible Differences Predictor
Bradley (1999)	IQ	Lum.		DWT (DB 9/7)		W	C	PS	E	Wavelet version of Daly (1993)
Lubin (1995)	IQ	Lum.		2DoG	yes	F,W	C	$L_{2,4}$	R	
Bolin and Meyer (1999)	IQ	Opp.		DWT (Haar)	yes	?	C	$L_{2,4}$	E	Simplified version of Lubin (1995)
Lubin and Fibush (1997)	VQ	$L^*u^*v^*$	yes	2DoG	yes	W	C(?)	L_p,H	R	Sarnoff JND (VQEG)
Lai and Kuo (2000)	IQ	Lum.		DWT (Haar)	yes	W	C(f,φ)	L_2	E	Wavelet-based metric
Teo and Heeger (1994a)	IQ	Lum.		steerable pyr.			C(φ)	L_2	E	Contrast gain control model
Lindh and van den Branden Lambrecht (1996)	VQ	Lum.		steerable pyr.		W	C(φ)	L_4	E	Video extension of above IQ metric
van den Branden Lambrecht (1996a)	VQ	Opp.		mod. Gabor		W	C	L_2	E	Color MPQM
D'Zmura et al. (1998)	IQ	AC_1C_2	?	Gabor	?	W	C(φ)		E	Color contrast gain control
Winkler (1998)	IQ	Opp.		steerable pyr.		W	C(φ)	L_2	R	See Sections 4.2 and 5.1
Winkler (1999b)	VQ	Opp.		steerable pyr.		W	C(φ)	L_2,L_4	R	See Sections 4.2 and 5.2 (VQEG)
Winkler (2000)	VQ	various		steerable pyr.		W	C(φ)	various	R	See Section 5.3
Masry and Hemami (2004)	VQ	Lum.		steerable pyr.		W	C(φ)	L_5,L_1	R	Low bitrate video, SSCQE data

Reference	Appl.	Colors	Lum.	Transform	CSF	(?)	Masking	Features	Pooling	Pooling	Valid.	Comments
Watson (1997)	IC	YC_BC_R	$L^?$	DCT					L_2		.	DCTune
Watson (1998), Watson et al. (1999)	VQ	YOZ		DCT	yes	W	C		$L_?$		R	DVQ metric (VQEG)
Wolf and Pinson (1999)	VQ	Lum.					C	Texture	L_2		R	Spatio-temporal blocks, 2 features
Tan et al. (1998)	VQ	Lum.				F		Edge	H, L_1	L_2	R	Cognitive emulator

?, not specified.

[1]IC, Image compression; IQ, Image quality; VQ, Video quality.

[2]Lum., Luminance; Opp., Opponent colors.

[3]γ, Monitor gamma; L, Luminance.

[4]2DoG, 2nd derivative of Gaussian; DB, Daubechies wavelet; DCT, Discrete Cosine Transform; DWT, Discrete Wavelet Transform; WHT, Walsh-Hadamard Transform.

[5]F, CSF filtering; W, CSF weighting.

[6]C, Contrast masking; C(f), ... over frequencies; C(φ), ... over orientations.

[7]H, Histogram; L_p, L_p-norm, exponent p; P_S, Probability summation.

[8]E, Examples; R, Subjective ratings.

contrast computation, and directional filtering, from which a contrast energy measure is calculated. It is subjected to a masking stage, which comprises a normalization process and a sigmoid nonlinearity. Finally, a distance measure or JND (just noticeable difference) map is computed as the L_p-norm of the masked responses. The VDM is one of the few models that take into account the eccentricity of the images in the observer's visual field. It was later modified to the Sarnoff JND metric for color video (Lubin and Fibush, 1997).

Another interesting distortion metric for still images was presented by Teo and Heeger (1994a,b). It is based on the response properties of neurons in the primary visual cortex and the psychophysics of spatial pattern detection. The model was inspired by analyses of the responses of single neurons in the visual cortex of the cat (Albrecht and Geisler, 1991; Heeger, 1992a,b), where a so-called *contrast gain control* mechanism keeps neural responses within the permissible dynamic range while at the same time retaining global pattern information (see section 4.2.4). In the metric, contrast gain control is realized by an excitatory nonlinearity that is inhibited divisively by a pool of responses from other neurons. The distortion measure is then computed from the resulting normalized responses by a simple squared-error norm. Contrast gain control models have become quite popular and have been generalized during recent years (Watson and Solomon, 1997; D'Zmura *et al.*, 1998; Graham and Sutter, 2000; Meese and Holmes, 2002).

Van den Branden Lambrecht (1996b) proposed a number of video quality metrics based on multi-channel vision models. The Moving Picture Quality Metric (MPQM) is based on a local contrast definition and Gabor-related filters for the spatial decomposition, two temporal mechanisms, as well as a spatio-temporal contrast sensitivity function and a simple intra-channel model of contrast masking (van den Branden Lambrecht and Verscheure, 1996). A color version of the MPQM based on an opponent color space was presented as well as a variety of applications and extensions of the MPQM (van den Branden Lambrecht, 1996a), for example, for assessing the quality of certain image features such as contours, textures, and blocking artifacts, or for the study of motion rendition (van den Branden Lambrecht *et al.*, 1999). Due to the MPQM's purely frequency-domain implementation of the spatio-temporal filtering process and the resulting huge memory requirements, it is not practical for measuring the quality of sequences with a duration of more than a few seconds, however. The Normalization Video Fidelity Metric (NVFM) by Lindh and van den Branden Lambrecht (1996) avoids this shortcoming by using a steerable pyramid transform for spatial filtering and discrete time-domain filter approximations of the temporal mechanisms. It is a spatio-temporal extension of Teo and Heeger's above-mentioned image distortion metric and implements inter-channel masking through an early model of contrast gain control. Both the MPQM and the NVFM are of particular relevance here because their implementations are used as the basis for the metrics presented in the following chapters of this book.

Recently, Masry and Hemami (2004) designed a metric for continuous video quality evaluation (CVQE) of low bitrate video. The metric works with luminance information only. It uses temporal filters and a wavelet transform for the perceptual decomposition, followed by CSF-weighting of the different bands, a gain control model, and pooling by means of two L_p-norms. Recursive temporal summation takes care of the low-pass nature of subjective quality ratings. The CVQE is one of the few vision-model based video quality metrics designed for and tested with low bitrate video.

3.4.4 Specialized Metrics

Metrics based on multi-channel vision models such as the ones presented above are the most general and potentially the most accurate ones (Winkler, 1999a). However, quality metrics need not necessarily rely on sophisticated general models of the human visual system; they can exploit a priori knowledge about the compression algorithm and the pertinent types of artifacts (see section 3.2) using ad hoc techniques or specialized vision models. While such metrics are not as versatile, they normally perform well in a given application area. Their main advantage lies in the fact that they often permit a computationally more efficient implementation. Since these artifact-based metrics are not the primary focus of this book, only a few are mentioned here.

One example of such specialized metrics is DCTune,[†] a method for optimizing JPEG image compression that was developed by Watson (1995, 1997). DCTune computes the JPEG quantization matrices that achieve the maximum compression for a specified perceptual distortion given a particular image and a particular set of viewing conditions. It considers visual masking by luminance and contrast techniques. DCTune can also compute the perceptual difference between two images.

Watson (1998) later extended the DCTune metric to video. In addition to the spatial sensitivity and masking effects considered in DCTune, this so-called Digital Video Quality (DVQ) metric relies on measurements of the visibility thresholds for temporally varying DCT quantization noise. It also models temporal forward masking effects by means of a masking sequence, which is

---

[†]A demonstration version of DCTune can be downloaded from http://vision.arc.nasa.gov/dctune/

produced by passing the reference through a temporal low-pass filter. A report of the DVQ metric's performance is given by Watson *et al.* (1999).

Wolf and Pinson (1999) developed another video quality metric (VQM) that uses reduced reference information in the form of low-level features extracted from spatio-temporal blocks of the sequences. These features were selected empirically from a number of candidates so as to yield the best correlation with subjective data. First, horizontal and vertical edge enhancement filters are applied to facilitate gradient computation in the feature extraction stage. The resulting sequences are divided into spatio-temporal blocks. A number of features measuring the amount and orientation of activity in each of these blocks are then computed from the spatial luminance gradient. To measure the distortion, the features from the reference and the distorted sequence are compared using a process similar to masking. This metric was one of the best performers in the latest VQEG FR-TV Phase II evaluation (see section 3.5.3).

Finally, Tan *et al.* (1998) presented a measurement tool for MPEG video quality. It first computes the perceptual impairment in each frame based on contrast sensitivity and masking with the help of spatial filtering and Sobel-operators, respectively. Then the PSNR of the masked error signal is calculated and normalized. The interesting part of this metric is its second stage, a cognitive emulator, that simulates higher-level aspects of perception. This includes the delay and temporal smoothing effect of observer responses, the nonlinear saturation of perceived quality, and the asymmetric behavior with respect to quality changes from bad to good and vice versa. This metric is one of the few models targeted at measuring the temporally varying quality of video sequences. While it still requires the reference as input, the cognitive emulator was shown to improve the predictions of subjective SSCQE MOS data.

3.5 METRIC EVALUATION

3.5.1 Performance Attributes

Quality as it is perceived by a panel of human observers (i.e. MOS) is the benchmark for any visual quality metric. There are a number of attributes that can be used to characterize a quality metric in terms of its prediction performance with respect to subjective ratings:[†]

[†]See the VQEG objective test plan at http://www.vqeg.org/ for details.

- *Accuracy* is the ability of a metric to predict subjective ratings with minimum average error and can be determined by means of the *Pearson linear correlation coefficient*; for a set of N data pairs (x_i, y_i), it is defined as follows:

$$r_P = \frac{\sum(x_i - \bar{x})(y_i - \bar{y})}{\sqrt{\sum(x_i - \bar{x})^2}\sqrt{\sum(y_i - \bar{y})^2}}, \quad (3.5)$$

where $\bar{x}$ and $\bar{y}$ are the means of the respective data sets. This assumes a linear relation between the data sets. If this is not the case, nonlinear correlation coefficients may be computed using equation (3.5) after applying a mapping function to one of the data sets, i.e. $\bar{y}_i = f(y_i)$. This helps to take into account saturation effects, for example. While nonlinear correlations are normally higher in absolute terms, the relations between them for different sets generally remain the same. Therefore, unless noted otherwise, only the linear correlations are used for analysis in this book, because our main interest lies in relative comparisons.

- *Monotonicity* measures if increases (decreases) in one variable are associated with increases (decreases) in the other variable, independently of the magnitude of the increase (decrease). Ideally, differences of a metric's rating between two sequences should always have the same sign as the differences between the corresponding subjective ratings. The degree of monotonicity can be quantified by the *Spearman rank-order correlation coefficient*, which is defined as follows:

$$r_S = \frac{\sum(\chi_i - \bar{\chi})(\gamma_i - \bar{\gamma})}{\sqrt{\sum(\chi - \bar{\chi})^2}\sqrt{\sum(\gamma_i - \bar{\gamma})^2}}, \quad (3.6)$$

where χ_i is the rank of x_i and γ_i is the rank of y_i in the ordered data series; $\bar{\chi}$ and $\bar{\gamma}$ are the respective midranks. The Spearman rank-order correlation is nonparametric, i.e. it makes no assumptions about the shape of the relationship between the x_i and y_i.

- The *consistency* of a metric's predictions can be evaluated by measuring the number of outliers. An outlier is defined as a data point (x_i, y_i) for which the prediction error is greater than a certain threshold, for example twice the standard deviation σ_{y_i} of the subjective rating differences for this data point, as proposed by VQEG (2000):

$$|x_i - y_i| > 2\sigma_{y_i}. \quad (3.7)$$

The *outlier ratio* is then simply defined as the number of outliers determined in this fashion in relation to the total number of data points:

$$r_O = N_O/N. \tag{3.8}$$

Evidently, the lower this outlier ratio, the better.

3.5.2 Metric Comparisons

While quality metric designs and implementations abound, only a handful of comparative studies exist that have investigated the prediction performance of metrics in relation to others.

Ahumada (1993) reviewed more than 30 visual discrimination models for still images from the application areas of image quality assessment, image compression, and halftoning. However, only a comparison table of the computational models is given; the performance of the metrics is not evaluated.

Comparisons of several *image* quality metrics with respect to their prediction performance were carried out by Fuhrmann *et al.* (1995), Jacobson (1995), Eriksson *et al.* (1998), Li *et al.* (1998), Martens and Meesters (1998), Mayache *et al.* (1998), and Avcibaş *et al.* (2002). These studies consider various pixel-based metrics as well as a number of single-channel and multi-channel models from the literature. Summarizing their findings and drawing overall conclusions is made difficult by the fact that test images, testing procedures, and applications differ greatly between studies. It can be noted that certain pixel-based metrics in the evaluations correlate quite well with subjective ratings for some test sets, especially for a given type of distortion or scene. They can be outperformed by vision-based metrics, where more complexity usually means more generality and accuracy. The observed gains are often so small, however, that the computational overhead does not seem justified.

Several measures of MPEG video quality were validated by Cermak *et al.* (1998). This comparison does not consider entire video quality metrics, but only a number of low-level features such as edge energy or motion energy and combinations thereof.

3.5.3 Video Quality Experts Group

The most ambitious performance evaluation of video quality metrics to date was undertaken by the Video Quality Experts Group (VQEG).[†] The group is composed of experts in the field of video quality assessment from industry, universities, and international organizations. VQEG was formed in 1997 with

[†]See http://www.vqeg.org/ for an overview of its activities.

the objective of collecting reliable subjective ratings for a well-defined set of test sequences and evaluating the performance of different video quality assessment systems with respect to these sequences.

In the first phase, the emphasis was on out-of-service testing (i.e. full-reference metrics) for production- and distribution-class video ('FR-TV'). Accordingly, the test conditions comprised mainly MPEG-2 encoded sequences with different profiles, different levels, and other parameter variations, including encoder concatenation, conversions between analog and digital video, and transmission errors. A set of 8-second scenes with different characteristics (e.g. spatial detail, color, motion) was selected by independent labs; the scenes were disclosed to the proponents only after the submission of their metrics. In total, 20 scenes were encoded for 16 test conditions each. Subjective ratings for these sequences were collected in large-scale experiments using the DSCQS method (see section 3.3.3). The VQEG test sequences and subjective experiments are described in more detail in sections 5.2.1 and 5.2.2.

The proponents of video quality metrics in this first phase were CPqD (Brazil), EPFL (Switzerland),[†] KDD (Japan), KPN Research/Swisscom (the Netherlands/Switzerland), NASA (USA), NHK/Mitsubishi (Japan), NTIA/ITS (USA), TAPESTRIES (EU), Technische Universität Braunschweig (Germany), and Tektronix/Sarnoff (USA).

The prediction performance of the metrics was evaluated with respect to the attributes listed in section 3.5.1. The statistical methods used for the analysis of these attributes were variance-weighted regression, nonlinear regression, Spearman rank-order correlation, and outlier ratio. The results of the data analysis showed that the performance of most models as well as PSNR are statistically equivalent for all four criteria, leading to the conclusion that no single model outperforms the others in all cases and for the entire range of test sequences (see also Figure 5.11). Furthermore, none of the metrics achieved an accuracy comparable to the agreement between different subject groups. The findings are described in detail in the final report (VQEG, 2000) and by Rohaly *et al.* (2000).

As a follow-up to this first phase, VQEG carried out a second round of tests for full-reference metrics ('FR-TV Phase II'); the final report was finished recently (VQEG, 2003). In order to obtain more discriminating results, this second phase was designed with a stronger focus on secondary distribution of digitally encoded television quality video and a wider range of distortions. New source sequences and test conditions were defined, and a

---

[†]This is the PDM described in section 4.2.

total of 128 test sequences were produced. Subjective ratings for these sequences were again collected using the DSCQS method. Unfortunately, the test sequences of the second phase are not public.

The proponents in this second phase were British Telecom (UK), Chiba University (Japan), CPqD (Brazil), NASA (USA), NTIA/ITS (USA), and Yonsei University (Korea). In contrast to the first phase, registration and calibration with the reference video had to be performed by each metric individually. Seven statistical criteria were defined to analyze the prediction performance of the metrics. These criteria all produced the same ranking of metrics, therefore only correlations are quoted here. The best metrics in the test achieved correlations as high as 94% with MOS, thus significantly outperforming PSNR, which had a correlation of about 70%. The results of this VQEG test are the basis for ITU-T Rec. J.144 (2004) and ITU-R Rec. BT.1683 (2004).

VQEG is currently working on an evaluation of reduced- and no-reference metrics for television ('RR/NR-TV'), for which results are expected by 2005, as well as an evaluation of metrics in a 'multimedia' scenario targeted at Internet and mobile video applications with the appropriate codecs, bitrates and frame sizes.

3.5.4 Limits of Prediction Performance

Perceived visual quality is an inherently subjective measure and can only be described statistically, i.e. by averaging over the opinions of a sufficiently large number of observers. Therefore the question is also how well subjects agree on the quality of a given image or video. In the first phase of VQEG tests, the correlations obtained between the average ratings of viewer groups from different labs are in the range of 90–95% for the most part (see Figure 3.11(a)). While the exact values certainly vary depending on the application and the quality range of the test set, this gives an indication of the limits on the prediction performance for video quality metrics. In the same study, the best-performing metrics only achieved correlations in the range of 80–85%, which is significantly lower than the inter-lab correspondences.

Nevertheless, it also becomes evident from Figure 3.11(b) that the DMOS values vary significantly between labs, especially for the low-quality test sequences, which was confirmed by an analysis of variance (ANOVA) carried out by VQEG (2000). The systematic offsets in DMOS observed between labs are quite small, but the slopes of the regression lines often deviate substantially from 1, which means that viewers in different labs had differing opinions about the quality range of the sequences (up to a factor

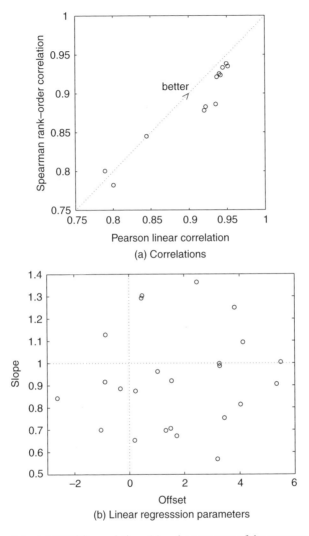

Figure 3.11 Inter-lab DMOS correlations (a) and parameters of the corresponding linear regressions (b).

of 2). On the other hand, the high inter-lab correlations indicate that ratings vary in a similar manner across labs and test conditions. In any case, the aim was to use the data from all subjects to compute global quality ratings for the various test conditions.

In the FR-TV Phase II tests (see section 3.5.3 above), a more rigorous test was used for studying the absolute performance limits of quality metrics. A statistically optimal model was defined on the basis of the subjective data to provide a quantitative upper limit on prediction performance (VQEG, 2003).

The assumption is that an optimal model would predict every MOS value exactly; however, the differences between the ratings of individual subjects for a given test clip cannot be predicted by an objective metric – it makes one prediction per clip, yet there are a number of different subjective ratings for that clip. These individual differences represent the residual variance of the optimal model, i.e. the minimum variance that can be achieved. For a given metric, the variance with respect to the individual subjective ratings is computed and compared against the residual variance of the optimal model using an F-test (see the VQEG final report for details). Despite the generally good performance of metrics in this test, none of the submitted metrics achieved a prediction performance that was statistically equivalent to the optimal model.

3.6 SUMMARY

The foundations of digital video and its visual quality were discussed. The major points of this chapter can be summarized as follows:

- Digital video systems are becoming increasingly widespread, be it in the form of digital TV and DVDs, in camcorders, on desktop computers or mobile devices. Guaranteeing a certain level of quality has thus become an important concern for content providers.
- Both analog and digital video coding standards exploit certain properties of the human visual system to reduce bandwidth and storage requirements. This compression as well as errors during transmission lead to artifacts and distortions affecting video quality.
- Subjective quality is a function of several different factors; it depends on the situation as well as the individual observer and can only be described statistically. Standardized testing procedures have been defined for gathering subjective quality data.
- Existing visual quality metrics were reviewed and compared. Pixel-based metrics such as MSE and PSNR are still popular despite their inability to reliably predict perceived quality across different scenes and distortion types. Many vision-based quality metrics have been developed that outperform PSNR. Nonetheless, no general-purpose metric has yet been found that is able to replace subjective testing.

With these facts in mind, we will now study vision models for quality metrics.

4

Models and Metrics

A theory has only the alternative of being right or wrong.
A model has a third possibility: it may be right, but irrelevant.

Manfred Eigen

Computational vision modeling is at the heart of this chapter. While the human visual system is extremely complex and many of its properties are still not well understood, models of human vision are the foundation for accurate general-purpose metrics of visual quality and have applications in many other fields of image processing. This chapter presents two concrete examples of vision models and quality metrics.

First, an isotropic measure of local contrast is described. It is based on the combination of directional analytic filters and is unique in that it permits the computation of an orientation- and phase-independent contrast for natural images. The design of the corresponding filters is discussed.

Second, a comprehensive perceptual distortion metric (PDM) for color images and color video is presented. It comprises several stages for modeling different aspects of the human visual system. Their design is explained in detail here. The underlying vision model is shown to achieve a very good fit to data from a variety of psychophysical experiments. A demonstration of the internal processing in this metric is also given.

Digital Video Quality - Vision Models and Metrics   Stefan Winkler
© 2005 John Wiley & Sons, Ltd   ISBN: 0-470-02404-6

4.1 ISOTROPIC CONTRAST

4.1.1 Contrast Definitions

As discussed in section 2.4.2, the response of the human visual system depends much less on the absolute luminance than on the relation of its local variations with respect to the surrounding luminance. This property is known as the *Weber–Fechner law*. Contrast is a measure of this relative variation of luminance.

Working with contrast instead of luminance can facilitate numerous image processing and analysis tasks. Unfortunately, a common definition of contrast suitable for all situations does not exist. This section reviews existing contrast definitions for artificial stimuli and presents a new isotropic measure of local contrast for natural images, which is computed from analytic filters (Winkler and Vandergheynst, 1999).

Mathematically, Weber's law can be formalized by *Weber contrast*:

$$C^W = \Delta L / L. \tag{4.1}$$

This definition is often used for stimuli consisting of small patches with a luminance offset ΔL on a uniform background of luminance L. In the case of sinusoids or other periodic patterns with symmetrical deviations ranging from L_{min} to L_{max}, which are also very popular in vision experiments, *Michelson contrast* (Michelson, 1927) is generally used:

$$C^M = \frac{L_{max} - L_{min}}{L_{max} + L_{min}}. \tag{4.2}$$

These two definitions are not equivalent and do not even share a common range of values: Michelson contrast can range from 0 to 1, whereas Weber contrast can range from to -1 to ∞. While they are good predictors of perceived contrast for simple stimuli, they fail when stimuli become more complex and cover a wider frequency range, for example Gabor patches (Peli, 1997). It is also evident that none of these simple global definitions is appropriate for measuring contrast in natural images. This is because a few very bright or very dark points would determine the contrast of the whole image, whereas actual human contrast perception varies with the *local average luminance*.

In order to address these issues, Peli (1990) proposed a *local band-limited contrast*:

$$C_j^P(x, y) = \frac{\psi_j * I(x, y)}{\phi_j * I(x, y)}, \tag{4.3}$$

where ψ_j is a band-pass filter at level j of a filter bank, and ϕ_j is the corresponding low-pass filter. An important point is that this contrast measure is well defined if certain conditions are imposed on the filter kernels. Assuming that the image and ϕ are positive real-valued integrable functions and ψ is integrable, $C_j^P(x, y)$ is a well defined quantity provided that the (essential) support of ψ is included in the (essential) support of ϕ. In this case $\phi_j * I(x, y) = 0$ implies $C_j^P(x, y) = 0$.

Using the band-pass filters of a pyramid transform, which can also be computed as the difference of two neighboring low-pass filters, equation (4.3) can be rewritten as

$$C_j^P(x, y) = \frac{(\phi_j - \phi_{j+1}) * I(x, y)}{\phi_{j+1} * I(x, y)} = \frac{\phi_j * I(x, y)}{\phi_{j+1} * I(x, y)} - 1. \qquad (4.4)$$

Lubin (1995) used the following modification of Peli's contrast definition in an image quality metric based on a multi-channel model of the human visual system:

$$C_j^L(x, y) = \frac{(\phi_j - \phi_{j+1}) * I(x, y)}{\phi_{j+2} * I(x, y)}. \qquad (4.5)$$

Here, the averaging low-pass filter has moved down one level. This particular local band-limited contrast definition has been found to be in good agreement with psychophysical contrast-matching experiments using Gabor patches (Peli, 1997).

The differences between C^P and C^L are most pronounced for higher-frequency bands. The lower one goes in frequency, the more spatially uniform the low-pass band in the denominator will become in both measures, finally approaching the overall luminance mean of the image. Peli's definition exhibits relatively high overshoots in certain image regions. This is mainly due to the spectral proximity of the band-pass and low-pass filters.

4.1.2 In-phase and Quadrature Mechanisms

Local contrast as defined above measures contrast only as incremental or decremental changes with respect to the local background. This is analogous to the symmetric (in-phase) responses of vision mechanisms. However, a complete description of contrast for complex stimuli has to include the anti-symmetric (quadrature) responses as well (Stromeyer and Klein, 1975; Daugman, 1985).

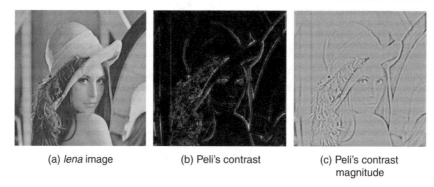

(a) *lena* image   (b) Peli's contrast   (c) Peli's contrast
                                                                     magnitude

Figure 4.1   Peli's local contrast from equation (4.3) and its magnitude computed for the
lena image.

This issue is demonstrated in Figure 4.1, which shows the contrast C^P
computed with an isotropic band-pass filter for the *lena* image. It can be
observed that C^P does not predict perceived contrast well due to its phase
dependence: C^P varies between positive and negative values of similar
amplitude at the border between bright and dark regions and exhibits zero-
crossings right where the perceived contrast is actually highest (note the
corresponding oscillations of the magnitude).

This behavior can be understood when C^P is computed for one-dimensional sinusoids with a constant C^M, as shown in Figure 4.2. The contrast
computed using only a symmetric filter actually oscillates between $\pm C^M$
with the same frequency as the underlying sinusoid, which is counter-
intuitive to the concept of contrast.

These examples underline the need for taking into account both the in-
phase and the quadrature component in order to be able to relate a generalized definition of contrast to the Michelson contrast of a sinusoidal grating.
Analytic filters represent an elegant way to achieve this: the magnitude of
the analytic filter response, which is the sum of the energy responses of
in-phase and quadrature components, exhibits the desired behavior in that it
gives a constant response to sinusoidal gratings. This is demonstrated in
Figure 4.2(c).

While the implementation of analytic filters in the one-dimensional case is
straightforward, the design of general two-dimensional analytic filters is less
obvious because of the difficulties involved when extending the Hilbert
transform to two dimensions (Stein and Weiss, 1971). This problem is
addressed in section 4.1.3 below.

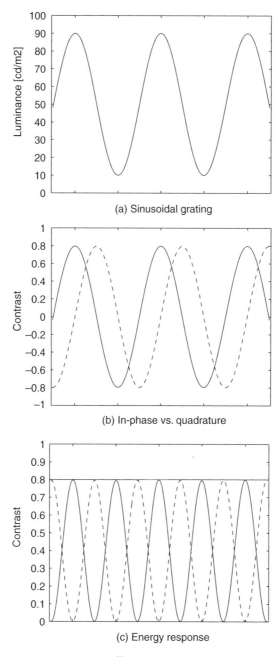

(a) Sinusoidal grating

(b) In-phase vs. quadrature

(c) Energy response

Figure 4.2 Sinusoidal grating with $C^M = 0.8$ (a). The contrast C^P computed using in-phase (solid) and quadrature (dashed) filters varies with the same frequency as the underlying sinusoid (b). Only the sum of the corresponding normalized energy responses is constant and equal to the grating's Michelson contrast (c).

Oriented measures of contrast can still be computed, because the Hilbert transform is well defined for filters whose angular support is smaller than π. Such contrast measures are useful for many image processing tasks. They can implement a multi-channel representation of low-level vision in accordance with the orientation selectivity of the human visual system and facilitate modeling aspects such as contrast sensitivity and pattern masking. They are in many vision models and their applications, for example in perceptual quality assessment of images and video (see sections 3.4.3 and 4.2). Contrast pyramids have also been found to reduce the dynamic range in the transform domain, which may find interesting applications in image compression (Vandergheynst and Gerek, 1999).

Lubin (1995), for example, applies oriented filtering to C_j^L from equation (4.5) and sums the squares of the in-phase and quadrature responses for each channel to obtain a phase-independent oriented measure of contrast energy. Using analytic orientation-selective filters $\eta_k(x, y)$, this oriented contrast can be expressed as

$$C_{jk}^L(x, y) = \left| \eta_k * C_j^L(x, y) \right|. \tag{4.6}$$

Alternatively, an oriented pyramid decomposition can be computed first, and contrast can be defined by normalizing the oriented sub-bands with a low-pass band:

$$C_{jk}^O(x, y) = \frac{\left| \psi_j * \eta_k * I(x, y) \right|}{\phi_{j+2} * I(x, y)} \tag{4.7}$$

Both of these approaches yield similar results in the decomposition of natural images. However, some noticeable differences occur around edges of high contrast.

4.1.3 Isotropic Local Contrast

The main problem in defining an isotropic contrast measure based on filtering operations is that if a flat response to a sinusoidal grating as with Michelson's definition is desired, 2-D analytic filters must be used. This requirement rules out the use of a single isotropic filter. As stated in the previous section, the main difficulty in designing 2-D analytic filters is the lack of a Hilbert transform in two dimensions. Instead, one must use the so-called *Riesz transforms* (Stein and Weiss, 1971), a series of transforms that are quite difficult to handle in practice.

In order to circumvent these problems, we describe an approach using a class of non-separable filters that generalize the properties of analytic functions in 2-D (Winkler and Vandergheynst, 1999). These filters are actually directional wavelets as defined by Antoine *et al.* (1999), which are square-integrable functions whose Fourier transform is strictly supported in a convex cone with the apex at the origin. It can be shown that these functions admit a holomorphic continuation in the domain $R^2 + jV$, where V is the cone defining the support of the function. This is a genuine generalization of the Paley–Wiener theorem for analytic functions in one dimension. Furthermore, if we require that these filters have a flat response to sinusoidal stimuli, it suffices to impose that the opening of the cone V be strictly smaller than π, as illustrated in Figure 4.3. This means that at least three such filters

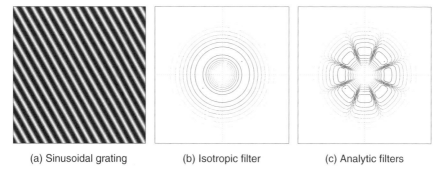

(a) Sinusoidal grating   (b) Isotropic filter   (c) Analytic filters

Figure 4.3 Computing the contrast of a two-dimensional sinusoidal grating (a): Using an isotropic band-pass filter, in-phase and quadrature components of the grating (dots) interfere within the same filter (b). This can be avoided using several analytic directional band-pass filters whose support covers an angle smaller than π (c).

are required to cover all possible orientations uniformly, but otherwise any number of filters is possible. Using a technique described below in section 4.1.4, such filters can be designed in a very simple and straightforward way; it is even possible to obtain dyadic oriented decompositions that can be implemented using a filter bank algorithm.

Working in polar coordinates (r, φ) in the Fourier domain, assume K directional wavelets $\hat{\Psi}(r, \varphi)$ satisfying the above requirements and

$$\sum_{k=0}^{K-1} \left| \hat{\Psi}(r, \varphi - 2\pi k/K) \right|^2 = \left| \hat{\psi}(r) \right|^2, \tag{4.8}$$

where $\hat{\psi}(r)$ is the Fourier transform of an isotropic dyadic wavelet, i.e.

$$\sum_{j=-\infty}^{\infty} \left| \hat{\psi}(2^j r) \right|^2 = 1 \tag{4.9}$$

and

$$\sum_{j=-J}^{\infty} \left| \hat{\psi}(2^j r) \right|^2 = \left| \hat{\phi}(2^J r) \right|^2. \tag{4.10}$$

where ϕ is the associated 2-D scaling function (Mallat and Zhong, 1992).

Now it is possible to construct an isotropic contrast measure C_j^I as the square root of the energy sum of these oriented filter responses, normalized as before by a low-pass band:

$$C_j^I(x, y) = \frac{\sqrt{2 \sum_k |\Psi_{jk} * I(x, y)|^2}}{\phi_j * I(x, y)}, \tag{4.11}$$

★here I is the input image, and Ψ_{jk} denotes the wavelet dilated by 2^{-j} and rotated by $2\pi k / K$. If the directional wavelet Ψ is in $L^1(\mathbb{R}^2) \cap L^2(\mathbb{R}^2)$, the convolution in the numerator of equation (4.11) is again a square-integrable function, and equation (4.8) shows that its L^2-norm is exactly what would have been obtained using the isotropic wavelet ψ. As can be seen in Figure 4.5, C_j^I is thus an orientation- and phase-independent quantity, but being defined by means of analytic filters it behaves as prescribed with respect to sinusoidal gratings (i.e. $C_j^I(x, y) \equiv C^M$ in this case).

Figure 4.4 shows an example of the pertinent decomposition for the *lena* image at three pyramid levels using $K = 8$ different orientations (the specific filters used in this example are described in section 4.1.4). The feature selection achieved by each directional filter is evident. The resulting isotropic contrast computed for the *lena* image at the three different levels is shown in Figure 4.5.

The figures clearly illustrate that C^I exhibits the desired omnidirectional and phase-independent properties. Comparing this contrast pyramid to the original image in Figure 4.1(a), it can be seen that the contrast features obtained with equation (4.11) correspond very well to the perceived contrast. Its localization properties obviously depend on the chosen pyramid level. The combination of the analytic oriented filter responses thus produces a

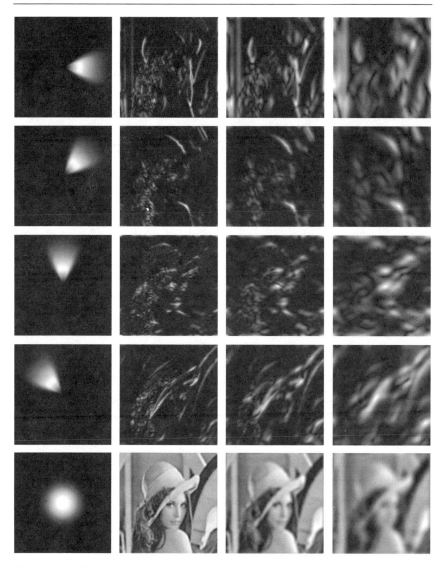

Figure 4.4 Filters used in the computation of isotropic local contrast (left column) and their responses for three different levels.

meaningful phase-independent measure of isotropic contrast. The examples show that it is a very natural measure of local contrast in an image. Isotropy is particularly important for applications where non-directional signals in an image are considered, e.g. spread-spectrum watermarking (Kutter and Winkler, 2002).

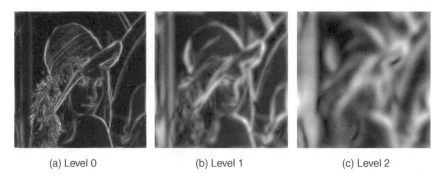

(a) Level 0   (b) Level 1   (c) Level 2

Figure 4.5 Three levels of isotropic local contrast $C_j^l(x, y)$ as given by equation (4.11) for the *lena* image.

4.1.4 Filter Design

As discussed in section 4.1.3, the computation of a robust isotropic contrast measure requires the use of a translation-invariant multi-resolution representation based on 2-D analytic filters. This can be achieved by designing a special Dyadic Wavelet Transform (DWT) using 2-D non-separable frames. The very weak design constraints of these frames permit the use of analytic wavelets, for which condition (4.8) can easily be fulfilled. This construction yields the following integrated wavelet packet (Vandergheynst *et al.*, 2000):

$$\left|\hat{\Gamma}(\vec{\omega})\right|^2 = \int_{1/2}^{1} \left|\hat{\psi}(a\vec{\omega})\right|^2 \frac{da}{a}. \tag{4.12}$$

Since the construction mainly works in the Fourier domain, it is very easy to add directional sensitivity by multiplying all Fourier transforms with a suitable angular window:

$$\hat{\Psi}(r, \varphi) = \hat{\Gamma}(r) \cdot \hat{\eta}(\varphi). \tag{4.13}$$

For this purpose, we introduce an infinitely differentiable, compactly supported function $\hat{\eta}(\varphi)$ such that

$$\sum_{k=0}^{K-1} |\hat{\eta}(\varphi - 2\pi k/K)|^2 = 1 \qquad \forall \varphi \in [0, 2\pi] \tag{4.14}$$

in order to satisfy condition (4.8).

This construction allows us to build oriented pyramids using a very wide class of dyadic wavelet decompositions. The properties of the filters involved in this decomposition can then be tailored to specific applications. The filters shown in Figure 4.5 are examples for $K = 8$ orientations.

The main drawback of this technique is the lack of fast algorithms. In particular, one would appreciate the existence of a pyramidal algorithm (Mallat, 1998), which is not guaranteed here because integrated wavelets and scaling functions are not necessarily related by a two-scale equation. On the other hand, it has been demonstrated that one can find quadrature filter approximations that achieve a fast implementation of the DWT while maintaining very accurate results (Gobbers and Vandergheynst, 2002; Muschietti and Torrésani, 1995). Once again, the advantage here is that it leaves us free to design our own dyadic frame.

In the examples presented above and in the applications proposed in other parts of this book, directional wavelet frames as described by Gobbers and Vandergheynst (2002) based on the *PLog* wavelet are used for the computation of isotropic local contrast according to equation (4.11). The *PLog* wavelet is defined as follows:

$$\psi_\tau(\vec{x}) = \frac{1}{\tau}\tilde{\psi}_\tau\left(\frac{\vec{x}}{\sqrt{\tau}}\right), \tag{4.15}$$

where

$$\tilde{\psi}_\tau(x,y) = \frac{(-1)^\tau}{2^{\tau-1}(\tau-1)!}\left(\frac{\partial^2}{\partial x^2}+\frac{\partial^2}{\partial y^2}\right)^\tau e^{-\frac{x^2+y^2}{2}}. \tag{4.16}$$

The integer parameter τ controls the number of vanishing moments and thus the shape of the wavelet. The filter response in the frequency domain broadens with decreasing τ. Several experiments were conducted to evaluate the impact of this parameter. The tests showed that values of $\tau > 2$ have to be avoided, because the filter selectivity becomes too low. Setting $\tau = 1$ has been found to be an appropriate value for our applications. The corresponding wavelet is also known as the *Log* wavelet or Mexican hat wavelet, i.e. the Laplacian of a Gaussian. Its frequency response is given by:

$$\hat{\psi}(r) = r^2\, e^{-\frac{r^2}{2}}. \tag{4.17}$$

For the directional separation of this isotropic wavelet, it is shaped in angular direction in the frequency domain:

$$\hat{\psi}_{jk}(r, \varphi) = \hat{\psi}_j(r) \cdot \hat{\eta}_k(\varphi). \tag{4.18}$$

The shaping function $\hat{\eta}_k(\varphi)$ used here is based on a combination of normalized Schwarz functions as defined by Gobbers and Vandergheynst (2002) that satisfies equation (4.14).

The number of filter orientations K is the parameter. The minimum number required by the analytic filter constraints, i.e. an angular support smaller than π, is three orientations. The human visual system emphasizes horizontal and vertical directions, so four orientations should be used as a practical minimum. To give additional weight to diagonal structures, eight orientations may be preferred (cf. Figure 4.4). Although using even more filters might result in a better analysis of the local neighborhood, our experiments indicate that there is no apparent improvement when using more than eight orientations, and the additional computational load outweighs potential benefits.

4.2 PERCEPTUAL DISTORTION METRIC

4.2.1 Metric Design

The perceptual distortion metric (PDM) is based on a contrast gain control model of the human visual system that incorporates spatial and temporal aspects of vision as well as color perception (Winkler, 1999b, 2000). It is based on a metric developed by Lindh and van den Branden Lambrecht (1996). The underlying vision model, an extension of a model for still images (Winkler, 1998), focuses on the following aspects of human vision:

- color perception, in particular the theory of opponent colors;
- the multi-channel representation of temporal and spatial mechanisms;
- spatio-temporal contrast sensitivity and pattern masking;
- the response properties of neurons in the primary visual cortex.

These visual aspects were already discussed in Chapter 2. Their implementation in the context of a perceptual distortion metric is explained in detail in the following sections.

A block diagram of the perceptual distortion metric is shown in Figure 4.6. The metric requires both the reference sequence and the distorted sequence

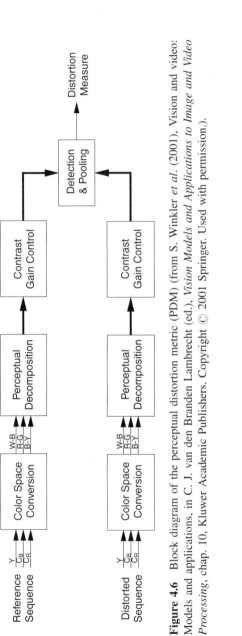

Figure 4.6 Block diagram of the perceptual distortion metric (PDM) (from S. Winkler *et al.* (2001), Vision and video: Models and applications, in C. J. van den Branden Lambrecht (ed.), *Vision Models and Applications to Image and Video Processing*, chap. 10, Kluwer Academic Publishers. Copyright © 2001 Springer. Used with permission.).

as inputs. After their conversion to the appropriate perceptual color space, each of the resulting three components is subjected to a spatio-temporal filter bank decomposition, yielding a number of perceptual channels. They are weighted according to contrast sensitivity data and subsequently undergo contrast gain control for pattern masking. Finally, the sensor differences are combined into a distortion measure.

4.2.2 Color Space Conversion

The color spaces used in many standards for coding visual information, e.g. PAL, NTSC, JPEG or MPEG, already take into account certain properties of the human visual system by coding nonlinear color difference components instead of linear *RGB* color primaries. Digital video is usually coded in $Y'C'_B C'_R$ space, where Y' encodes luminance, C'_B the difference between the blue primary and luminance, and C'_R the difference between the red primary and luminance. The PDM on the other hand relies on the theory of opponent colors for color processing, which states that the color information received by the cones is encoded as white-black, red-green and blue-yellow color difference signals (see section 2.5.2).

Conversion from $Y'C'_B C'_R$ to opponent color space requires a series of transformations as illustrated in Figure 4.7. $Y'C'_B C'_R$ color space is defined in

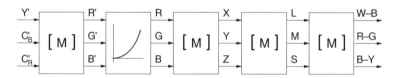

Figure 4.7   Color space conversion from component video $Y'C'_B C'_R$ to opponent color space.

ITU-R Rec. BT.601-5. Using 8 bits for each component, Y' is coded with an offset of 16 and an amplitude range of 219, while C'_B and C'_R are coded with an offset of 128 and an amplitude range of ±112. The extremes of the coding range are reserved for synchronization and signal processing headroom, which requires clipping prior to conversion. Nonlinear $R'G'B'$ values in the range [0,1] are then computed from 8-bit $Y'C'_B C'_R$ as follows (Poynton, 1996):

$$\begin{bmatrix} R' \\ G' \\ B' \end{bmatrix} = \frac{1}{219} \begin{bmatrix} 1 & 0 & 1.371 \\ 1 & -0.336 & -0.698 \\ 1 & 1.732 & 0 \end{bmatrix} \cdot \left(\begin{bmatrix} Y' \\ C'_B \\ C'_R \end{bmatrix} - \begin{bmatrix} 16 \\ 128 \\ 128 \end{bmatrix} \right). \quad (4.19)$$

Each of the resulting three components undergoes a power-law nonlinearity of the form x^γ with $\gamma \approx 2.5$ to produce linear *RGB* values. This is required to counter the gamma correction used in nonlinear $R'G'B'$ space to compensate for the behavior of a conventional CRT display (cf. section 3.1.1).

RGB space further assumes a particular display device, or to be more exact, a particular spectral power distribution of the light emitted from the display phosphors. Once the phosphor spectra of the monitor of interest have been determined, the device-independent CIE *XYZ* tristimulus values can be calculated. The primaries of contemporary monitors are closely approximated by the following transformation defined in ITU-R Rec. BT.709-5 (2002):

$$\begin{bmatrix} X \\ Y \\ Z \end{bmatrix} = \begin{bmatrix} 0.412 & 0.358 & 0.180 \\ 0.213 & 0.715 & 0.072 \\ 0.019 & 0.119 & 0.950 \end{bmatrix} \cdot \begin{bmatrix} R \\ G \\ B \end{bmatrix}. \tag{4.20}$$

The CIE *XYZ* tristimulus values form the basis for conversion to an HVS-related color space. First, the responses of the L-, M-, and S-cones on the human retina (see section 2.2.1) are computed as follows (Hunt, 1995):

$$\begin{bmatrix} L \\ M \\ S \end{bmatrix} = \begin{bmatrix} 0.240 & 0.854 & -0.044 \\ -0.389 & 1.160 & 0.085 \\ -0.001 & 0.002 & 0.573 \end{bmatrix} \cdot \begin{bmatrix} X \\ Y \\ Z \end{bmatrix}. \tag{4.21}$$

The *LMS* values can now be converted to an opponent color space. A variety of opponent color spaces have been proposed, which use different ways to combine the cone responses. The PDM relies on a recent opponent color model by Poirson and Wandell (1993, 1996). This particular opponent color space has been designed for maximum pattern-color separability, which has the advantage that color perception and pattern sensitivity can be decoupled and treated in separate stages in the metric. The spectral sensitivities of its W-B, R-G and B-Y components are shown in Figure 2.14. These components are computed from *LMS* values via the following transformation (Poirson and Wandell, 1993):

$$\begin{bmatrix} W-B \\ R-G \\ B-Y \end{bmatrix} = \begin{bmatrix} 0.990 & -0.106 & -0.094 \\ -0.669 & 0.742 & -0.027 \\ -0.212 & -0.354 & 0.911 \end{bmatrix} \cdot \begin{bmatrix} L \\ M \\ S \end{bmatrix}. \tag{4.22}$$

4.2.3 Perceptual Decomposition

As discussed in sections 2.3.2 and 2.7, many cells in the human visual system are selectively sensitive to certain types of signals, such as patterns of a particular frequency or orientation. This multi-channel theory of vision has proven successful in explaining a wide variety of perceptual phenomena. Therefore, the PDM implements a decomposition of the input into a number of channels based on the spatio-temporal mechanisms in the visual system. This perceptual decomposition is performed first in the temporal and then in the spatial domain. As discussed in section 2.4.2, this separation is not entirely unproblematic, but it greatly facilitates the implementation of the decomposition. Besides, these two domains can be consolidated in the fitting process as described in section 4.2.6.

4.2.3.1 Temporal Mechanisms

The characteristics of the temporal mechanisms in the human visual system were described in section 2.7.2. The temporal filters used in the PDM are based on the work by Fredericksen and Hess (1997, 1998), who model temporal mechanisms using derivatives of the following impulse response function:

$$h(t) = e^{-\left(\frac{\ln(t/\tau)}{\sigma}\right)^2}.$$

(4.23)

They achieve a very good fit to their experimental data using only this function and its second derivative, corresponding to one sustained and one transient mechanism, respectively. For a typical choice of parameters $\tau = 160\,\text{ms}$ and $\sigma = 0.2$, the frequency responses of the two mechanisms are shown in Figure 4.8(a), and the corresponding impulse responses are shown in Figure 4.8(b).

For use in the PDM, the temporal mechanisms have to be approximated by digital filters. The primary design goal for these filters is to keep the delay to a minimum, because in some applications of distortion metrics such as monitoring and control, a short response time is crucial. This fact together with limitations of memory and computing power favor time-domain implementations of the temporal filters over frequency-domain implementations. A trade-off has to be found between an acceptable delay and the accuracy with which the temporal mechanisms ought to be approximated.

Two digital filter types are investigated for modeling the temporal mechanisms, namely recursive infinite impulse response (IIR) filters and

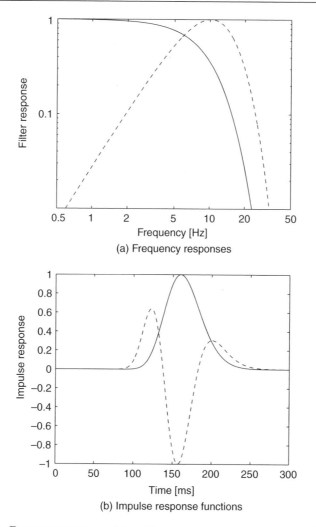

(a) Frequency responses

(b) Impulse response functions

Figure 4.8 Frequency responses (a) and impulse response functions (b) of sustained (solid) and transient (dashed) mechanisms of vision (Fredericksen and Hess, 1997, 1998).

nonrecursive finite impulse response (FIR) filters with linear phase. The filters are computed by means of a least-squares fit to the normalized frequency magnitude response of the corresponding mechanism as given by the Fourier transforms of $h(t)$ and $h''(t)$ from equation (4.23).

· Figures 4.9 and 4.10 show the resulting IIR and FIR filter approximations for a sampling frequency of 50 Hz. Excellent fits to the frequency

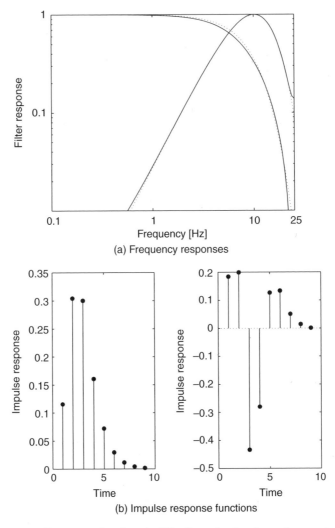

Figure 4.9 IIR filter approximations (solid) of sustained and transient mechanisms of vision (dotted) for a sampling frequency of 50 Hz.

responses are obtained with both filter types. An IIR filter with 2 poles and 2 zeros is fitted to the sustained mechanism, and an IIR filter with 5 poles and 5 zeros is fitted to the transient mechanism. For FIR filters, a filter length of 9 taps is entirely sufficient for both mechanisms. These settings have been found to yield acceptable delays while maintaining a good approximation of the temporal mechanisms.

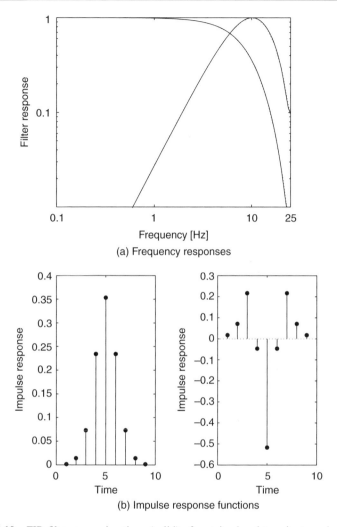

Figure 4.10 FIR filter approximations (solid) of sustained and transient mechanisms of vision (dotted) for a sampling frequency of 50 Hz.

The impulse responses of the IIR and FIR filters are shown in Figures 4.9(b) and 4.10(b), respectively. It can be seen that all of them are nearly zero after 7 to 8 time samples. For television frame rates, this corresponds to a delay of approximately 150 ms in the metric. Due to the symmetry restrictions imposed on the impulse response of linear-phase FIR filters, their approximation of the impulse response cannot be as good as with IIR filters.

On the other hand, linear phase can be important for video processing applications, as the delay introduced is the same for all frequencies.

In the present implementation, the temporal low-pass filter is applied to all three color channels, while the band-pass filter is applied only to the luminance channel in order to reduce computing time. This simplification is based on the fact that our sensitivity to color contrast is reduced for high frequencies (see section 2.4.2).

4.2.3.2 Spatial Mechanisms

The characteristics of the spatial mechanisms in the human visual system were discussed in section 2.7.1. Given the bandwidths mentioned there, and considering the decrease in contrast sensitivity at high spatial frequencies (see section 2.4.2), the spatial frequency plane for the achromatic channel can be covered by 4–6 spatial frequency-selective and 4–8 orientation-selective mechanisms. A further reduction of orientation selectivity can affect modeling accuracy, as was reported in a comparison of two models with 3 and 6 orientation-selective mechanisms (Teo and Heeger, 1994a,b).

Taking into account the larger orientation bandwidths of the chromatic channels, 2–3 orientation-selective mechanisms may suffice there. Chromatic sensitivity remains high down to very low spatial frequencies, which necessitates a low-pass mechanism and possibly additional spatial frequency-selective mechanisms at this end. For reasons of implementation simplicity, the same decomposition filters are used for chromatic and achromatic channels.

Many different filters have been proposed as approximations to the multi-channel representation of visual information in the human visual system. These include Gabor filters, the cortex transform (Watson, 1987a), and wavelets. We have found that the exact shape of the filters is not of paramount importance, but our goal here is also to obtain a good trade-off between implementation complexity, flexibility, and prediction accuracy.

In the PDM, therefore, the decomposition in the spatial domain is carried out by means of the steerable pyramid transform proposed by Simoncelli *et al.* (1992).[†] This transform decomposes an image into a number of spatial frequency and orientation bands. Its basis functions are directional derivative operators. For use within a vision model, the steerable pyramid transform has the advantage of being rotation-invariant and self-inverting while minimizing

---

[†]The source code for the steerable pyramid transform is available at http://www.cis.upenn.edu/~eero/steerpyr.html

the amount of aliasing in the sub-bands. In the present implementation, the basis filters have octave bandwidth and octave spacing. Five sub-band levels with four orientation bands each plus one low-pass band are computed; the bands at each level are tuned to orientations of 0, 45, 90 and 135 degrees (Figure 4.11). The same decomposition is used for the W-B, R-G and B-Y channels.

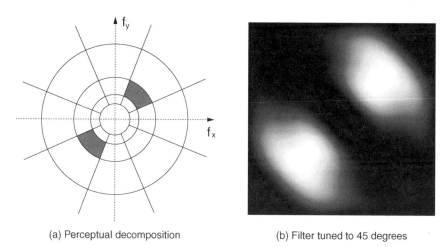

| (a) Perceptual decomposition | (b) Filter tuned to 45 degrees |

Figure 4.11 Illustration of the partitioning of the spatial frequency plane by the steerable pyramid transform (Simoncelli *et al.*, 1992). Three levels plus one (isotropic) low-pass filter are shown (a). The shaded region indicates the spectral support of a single sub-band, whose actual frequency response is plotted (b) (from S. Winkler *et al.* (2001), Vision and video: Models and applications, in C. J. van den Branden Lambrecht (ed.), *Vision Models and Applications to Image and Video Processing*, chap. 10, Kluwer Academic Publishers. Copyright © 2001 Springer. Used with permission.).

4.2.3.3 Contrast Sensitivity

After the temporal and spatial decomposition, each channel is weighted such that the ensemble of all filters approximates the spatio-temporal contrast sensitivity of the human visual system. While this approach is less accurate than pre-filtering the W-B, R-G and B-Y channels with their respective contrast sensitivity functions, it is easier to implement and saves computing time. The resulting approximation accuracy is still very good, as will be shown in section 4.2.6.

4.2.4 Contrast Gain Control

Modeling pattern masking is one of the most critical components of video quality assessment because the visibility of distortions is highly dependent on the local background. As discussed in section 2.6.1, masking occurs when a stimulus that is visible by itself cannot be detected due to the presence of another. Within the framework of quality assessment it is helpful to think of the distortion or the coding noise as being masked by the original image or sequence acting as background. Masking explains why similar coding artifacts are disturbing in certain regions of an image while they are hardly noticeable in others.

Masking is strongest between stimuli located in the same perceptual channel, and many vision models are limited to this intra-channel masking. However, psychophysical experiments show that masking also occurs between channels of different orientations (Foley, 1994), between channels of different spatial frequency, and between chrominance and luminance channels (Switkes et al., 1988; Cole et al., 1990; Losada and Mullen, 1994), albeit to a lesser extent.

Models have been proposed which explain a wide variety of empirical contrast masking data within a process of contrast gain control. These models were inspired by analyses of the responses of single neurons in the visual cortex of the cat (Albrecht and Geisler, 1991; Heeger, 1992a,b), where contrast gain control serves as a mechanism to keep neural responses within the permissible dynamic range while at the same time retaining global pattern information.

Contrast gain control can be modeled by an excitatory nonlinearity that is inhibited divisively by a pool of responses from other neurons. Masking occurs through the inhibitory effect of the normalizing pool (Foley, 1994; Teo and Heeger, 1994a). Watson and Solomon (1997) presented an elegant generalization of these models that facilitates the integration of many kinds of channel interactions as well as spatial pooling. Introduced for luminance images, this contrast gain control model is now extended to color and to sequences as follows: let $a = a(t, c, f, \varphi, x, y)$ be a coefficient of the perceptual decomposition in temporal channel t, color channel c, frequency band f, orientation band φ, at location x, y. Then the corresponding sensor output $s = s(t, c, f, \varphi, x, y)$ is computed as

$$s = k \frac{a^p}{b^2 + h * a^q}. \tag{4.24}$$

The excitatory path in the numerator consists of a power-law nonlinearity with exponent p. Its gain is controlled by the inhibitory path in the denominator, which comprises a nonlinearity with a possibly different exponent q and a saturation constant b to prevent division by zero. The

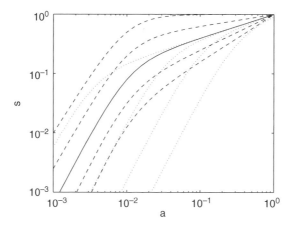

Figure 4.12 Illustration of contrast gain control as given by equation (4.24). The sensor output s is plotted as a function of the normalized input a for $q = 2$, $k = 1$, and no pooling. Solid line: $p = 2.4$, $b^2 = 10^{-4}$. Dashed lines from left to right: $p = 2.0$, $2.2, 2.6, 2.8$. Dotted lines from left to right: $b^2 = 10^{-5}, 10^{-3}, 10^{-2}, 10^{-1}$.

factor k is used to adjust the overall gain of the mechanism. The effects of these parameters are visualized in Figure 4.12.

In the implementation of Teo and Heeger (1994a,b), which is based on a direct model of neural cell responses (Heeger, 1992b), the exponents of both the excitatory and inhibitory nonlinearity are fixed at $p = q = 2$ so as to be able to work with local energy measures. However, this procedure rapidly saturates the sensor outputs (see top curve in Figure 4.12), which necessitates multiple contrast bands (i.e. several different k's and b's) for all coefficients in order to cover the full range of contrasts. Watson and Solomon (1997) showed that the same effect can be achieved with a single contrast band when $p > q$. This approach reduces the number of model parameters considerably and simplifies the fitting process, which is why it is used in the PDM. The fitting procedure for the contrast gain control stage and its results are discussed in more detail in section 4.2.6 below.

In the inhibitory path, filter responses are pooled over different channels by means of a convolution with the pooling function $h = h(t, c, f, \varphi, x, y)$. In its most general form, the pooling operation in the inhibitory path may combine coefficients from the dimensions of time, color, temporal frequency, spatial frequency, orientation, space, and phase. In the present implementation of the distortion metric, it is limited to orientation. A Gaussian pooling kernel is used for the orientation dimension as a first approximation to channel interactions.

4.2.5 Detection and Pooling

It is believed that the information represented in various channels within the primary visual cortex is integrated in the subsequent brain areas. This process can be simulated by gathering the data from these channels according to rules of probability or vector summation, also known as pooling. However, little is known about the nature of the actual integration taking place in the brain. There is no firm experimental evidence that the mathematical assumptions and equations presented below are a good description of the pooling mechanism in the human visual system (Quick, 1974; Fredericksen and Hess, 1998; Meese and Williams, 2000).

If there are a number of independent 'reasons' i for an observer noticing the presence of a distortion, each having probability P_i respectively, the overall probability P of the observer noticing the distortion is

$$P = 1 - \prod_i (1 - P_i). \tag{4.25}$$

This is the probability summation rule. The dependence of P_i on the distortion strength x_i can be described by the psychometric function

$$P_i = 1 - e^{-x_i^{\beta_i}}. \tag{4.26}$$

This is one version of a distribution function studied by Weibull (1951) and first applied to vision by Quick (1974). β determines the slope of the function. Under the homogeneity assumption that all β_i are equal (Nachmias, 1981), equations (4.25) and (4.26) can be combined to yield

$$P_i = 1 - e^{-\sum x_i^{\beta}}. \tag{4.27}$$

The sum in the exponent of this equation is in itself an indicator of the visibility of distortions. Therefore, models may postulate a combination of mechanism responses before producing an estimate of detection probability. This is referred to as vector summation or Minkowski summation:

$$x = \sum_i x_i^{\beta}. \tag{4.28}$$

This principle is also applied in the PDM. Its detection and pooling stage combines the elementary differences between N sensor outputs of the contrast gain control stage for the reference sequence $s = s(t, c, f, \varphi, x, y)$

and the distorted sequence $\tilde{s} = \tilde{s}(t, c, f, \varphi, x, y)$ over several dimensions by means of a Minkowski distance:

$$\Delta = \sqrt[\beta]{\frac{1}{N} \sum |s - \tilde{s}|^{\beta}}. \qquad (4.29)$$

Often this summation is carried out over all dimensions in order to obtain a single distortion rating for an image or sequence, but in principle, any subset of dimensions can be used, depending on what kind of result is desired. For example, pooling over pixel locations may be omitted to produce a distortion map for every frame of the sequence (examples are shown in section 4.2.7 below). The combination may be nested as well: pooling can be limited to single frames first to determine the variation of distortions over time, and the total distortion can be computed from the values for each frame.

4.2.6 Parameter Fitting

The model contains several parameters that have to be adjusted in order to accurately represent the human visual system (see Figure 4.13). Threshold data from contrast sensitivity and contrast masking experiments are used for this procedure. In the fitting process, the inputs to the metric imitate the stimuli used in these experiments, and the free model parameters are adjusted in such a way that the metric approximates these threshold curves by determining the stimulus strengths for which the output of the metric remains at a given constant.

Contrast sensitivity is modeled by setting the gains of the spatial and temporal filters in such a way that the model predictions match empirical threshold data from spatio-temporal contrast sensitivity experiments for both color and luminance stimuli. For the W-B channels, the weights are chosen so as to match contrast sensitivity data from Kelly (1979a,b). For the R-G and B-Y channels, similar data from Mullen (1985) or Kelly (1983) are used. As an example, the fit to contrast sensitivity data for blue-yellow gratings is shown in Figure 4.14(a). The individual decomposition filters used in the approximation by the model can be clearly distinguished. The parameters obtained in this manner for the sustained (low-pass) and transient (band-pass) mechanisms are listed in Table 4.1 for a typical television viewing setup.

The parameters k, p and b of the contrast gain control stage from equation (4.24) are determined by fitting the model's responses to masked gratings; the inhibitory exponent is fixed at $q = 2$ in this implementation, as it is mainly the difference $p - q$ which matters (Watson and Solomon, 1997). For

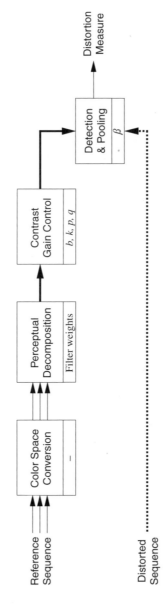

Figure 4.13 Free model parameters in the different stages of the PDM.

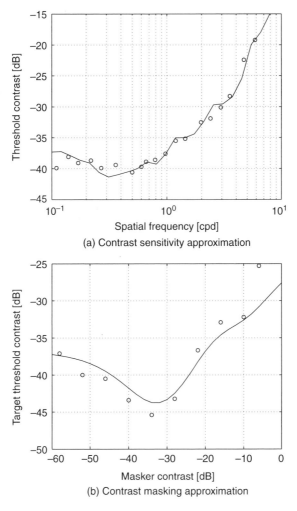

Figure 4.14 Model approximations (solid curves) of psychophysical data (dots). (a) Contrast sensitivity data for blue-yellow gratings from Mullen (1985). (b) Contrast masking data for red-green gratings from Switkes *et al.* (1988).

Table 4.1 Filter weights

Level	0	1	2	3	4
W-B, LP	5.0	19.2	139.5	478.6	496.5
W-B, BP	112.8	141.0	179.4	205.7	120.0
R-G, LP	154.2	354.0	404.0	184.6	27.0
B-Y, LP	125.6	332.7	381.4	131.5	28.6

the W-B channel, empirical data from several intra- and inter-channel contrast masking experiments conducted by Foley (1994) are used. For the R-G and B-Y channels, the parameters are adjusted to fit similar data presented by Switkes *et al.* (1988), as shown in Figure 4.14(b) for the R-G channel. The parameters obtained in this manner for all three color channels are listed in Table 4.2 for a typical television viewing setup.

Table 4.2   Contrast gain control parameters

	b	k	p	q
W-B	6.968	0.29778	2.1158	2
R-G	21.904	0.11379	2.3447	2
B-Y	13.035	0.07712	2.2788	2

The choice of the exponent β in the pooling stage is less obvious. Different exponents have been found to yield good results for different experiments and implementations. $\beta = 2$ corresponds to the ideal observer formalism under independent Gaussian noise, which assumes that the observer has complete knowledge of the stimuli and uses a matched filter for detection. The sensor outputs can be considered as the mean values of noisy sensors. Assuming an additive, independent, identically distributed Gaussian noise with zero mean and a standard deviation independent of the sensor outputs, a squared-error norm detection stage gives the probability that the ideal observer detects the distortion (Teo and Heeger, 1994a). In a study of subjective experiments with coding artifacts, $\beta \approx 2$ yielded the best results (de Ridder, 1992). Intuitively, a few strong distortions may draw the viewer's attention more than many weak ones. This behavior can be emphasized with larger exponents. In the PDM, pooling over channels and over pixels is carried out with $\beta = 2$, whereas $\beta = 4$ is used for pooling over frames. This combination was found to give good results in the fitting process.

The fitting results shown in Figures 4.14(a) and 4.14(b) demonstrate that the overall quality of the fits to the above-mentioned empirical data is quite good and close to the difference between measurements from different observers. Most of the effects found in the psychophysical experiments are captured by the model. However, two drawbacks of this modeling approach should be noted. Because of the nonlinear nature of the model, the parameters can only be determined by means of an iterative least-squares fitting process, which is computationally intensive. Furthermore, the model is not very flexible: once a good set of parameters has been found, it is only valid for a particular viewing setup (i.e. viewing distance and resolution).

4.2.7 Demonstration

The basketball sequence is used to briefly demonstrate the internal processing of the proposed distortion metric. This sequence contains a lot of spatial detail, a considerable amount of fast motion (the players in the foreground), and slow camera panning, which makes it an interesting sequence for a spatio-temporal model.

The frame size of the sequence is 704×576 pixels. It was encoded at a bitrate of 4 Mb/s with the MPEG-2 encoder of the MPEG Software Simulation Group.[†] A sample frame, its encoded counterpart, and the pixel-wise difference between them are shown in Figure 4.15. The W-B, R-G and B-Y components resulting from the conversion to opponent color space are shown in Figure 4.16. Note the emphasis of the ball in the R-G channel as well as the yellow curved line on the floor in the B-Y channel. The W-B component

(a) Reference frame   (b) Encoded frame   (c) Frame difference

Figure 4.15 Sample frame from the basketball sequence. The reference, its encoded counterpart, and the pixel-wise difference between them are shown.

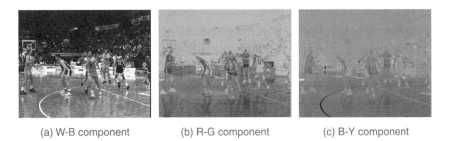

(a) W-B component   (b) R-G component   (c) B-Y component

Figure 4.16 The W-B, R-G and B-Y components resulting from the conversion to opponent color space.

[†]The source code is available at http://www.mpeg.org/~tristan/MPEG/MSSG/

looks different from the gray-level image in Figure 4.15 because the transform coefficients differ and because of the gamma-correcting nonlinearity that has been applied as part of the color space conversion.

The color space conversions are followed by the perceptual decomposition. The results of applying the temporal low-pass and band-pass filters to the W-B channel are shown in Figure 4.17. As can be seen, the ball virtually

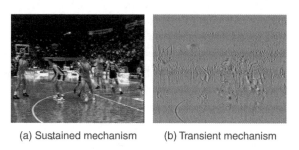

(a) Sustained mechanism   (b) Transient mechanism

Figure 4.17   The temporally low-pass and band-pass filtered W-B channels.

disappears in the low-pass channel, while it is clearly visible in the band-pass channel. As mentioned before, the R-G and B-Y channels are subjected only to the low-pass filter. The decomposition in the spatial domain increases the total number of channels even further; only a small selection is shown in Figure 4.18, namely the first, third and fifth level of the pyramid at an orientation of 45° constructed from the low-pass filtered W-B channel. The images are downsampled in the pyramid transform and have been upsampled

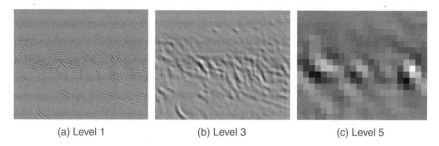

(a) Level 1   (b) Level 3   (c) Level 5

Figure 4.18   Three levels at an orientation of 45° of the pyramid constructed from the low-pass filtered W-B channel.

to their original size in the figure. They show very well how different features are emphasized in the different sub-bands, for example the lines on the floor in the high-frequency channel, the players leaning to the left in the medium-frequency channel, and the barricades around the field in the low-frequency channel.

Figure 4.19 shows the output of the PDM as separate distortion maps for each color and temporal channel. Note that these distortion maps also include

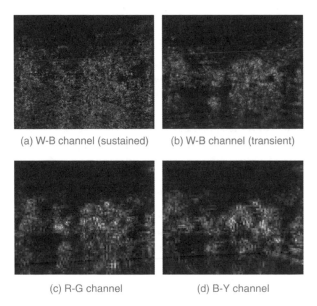

(a) W-B channel (sustained)   (b) W-B channel (transient)

(c) R-G channel   (d) B-Y channel

Figure 4.19 Distortion maps of the sample frame for the low-pass and band-pass W-B channels, the R-G channel and the B-Y channel. The images are normalized to better show the spatial structure; the absolute distortion values in the color channels are much smaller than in the W-B channels.

temporal aspects of the distortions, i.e. they depend on the neighboring frames. It is evident that all four distortion maps are very different from the simple pixel-wise difference between the reference frame and the encoded frame shown in Figure 4.15. Most of the visible artifacts appear in the W-B band-pass channel around the silhouettes of the players currently in motion. The distortions in the color channels are small compared to the other channels, but they have been normalized in the figures to reveal more spatial detail. Note that the distortions in the R-G and B-Y channels show a distinct block structure. This is due to the subsampling in the pyramid transform and shows that the model correctly emphasizes low-frequency distortions in the color channels. Compared to the pixel-wise frame difference shown in Figure 4.15, much less weight is given to the distortions in the top half of the frame, where they are masked by the high spatial detail. Instead, the distortions of the well-defined players moving on the relatively uniform playing field are emphasized, which is in good agreement with human visual perception.

4.3 SUMMARY

Two models of different vision aspects were presented in this chapter:

- An isotropic local contrast measure was constructed from the combination of analytic directional filter responses. The proposed definition is the first omnidirectional, phase-independent measure of local contrast that can be applied to natural images and corresponds very well to perceived contrast.
- A perceptual distortion metric (PDM) for digital color video was described. It is based on a model of the human visual system, whose design and components were discussed. The model takes into account color perception, the multi-channel architecture of temporal and spatial mechanisms, spatio-temporal contrast sensitivity, pattern masking and channel interactions. The PDM was shown to accurately fit data from psychophysical experiments on contrast sensitivity and pattern masking. The metric's output is consistent with human observation.

The performance of the PDM will now be analyzed by means of extensive data from subjective experiments using natural images and sequences in Chapter 5. The isotropic contrast will be combined with the PDM in section 6.3 in the form of a sharpness measure to improve the accuracy of the metric's predictions.

5

Metric Evaluation

I have had my results for a long time,
but I do not yet know how I am to arrive at them.

Carl Friedrich Gauss

Subjective experiments are necessary in order to evaluate models of human vision, and subjective ratings form the benchmark for visual quality metrics. In this chapter, the perceptual distortion metric (PDM) introduced in Chapter 4 is evaluated with the help of data from subjective experiments with natural images and video. The test images and sequences as well as the experimental procedures are presented, and the performance of the metric is discussed.

First the PDM is validated with respect to threshold data from natural images. The remainder of this chapter is then devoted to analyses based on data obtained in the framework of the Video Quality Experts Group (VQEG, 2000). The prediction performance of the PDM for numerous test sets is analyzed in comparison to subjective ratings and to competing metrics. Finally, various implementation choices for the different stages of the PDM are evaluated, in particular the choice of the color space, the decomposition filters, and the pooling algorithm.

5.1 STILL IMAGES

5.1.1 Test Images

The database used for the validation of the PDM with respect to still images was generously provided by van den Branden Lambrecht and Farrell (1996).

Digital Video Quality - Vision Models and Metrics   Stefan Winkler
© 2005 John Wiley & Sons, Ltd   ISBN: 0-470-02404-6

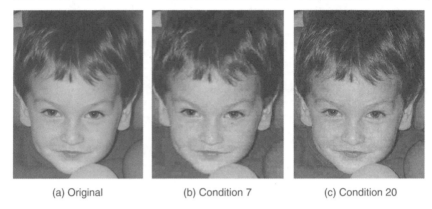

(a) Original   (b) Condition 7   (c) Condition 20

Figure 5.1   Original test image and two examples of distorted versions.

It consists of distorted versions of a color image of 320×400 pixels in size, showing the face of a child surrounded by colorful balls (see Figure 5.1(a)). To create the test images, the original was JPEG-encoded, and the coding noise was determined in YUV space by computing the difference between the original and the compressed image. Subsequently, the coding noise was scaled by a factor ranging from -1 to 1 in the Y, U, and V channel separately and was then added back to the original in order to obtain the distorted images. A total of 20 test conditions were defined, which are listed in Table 5.1, and the test series were created by varying the noise intensity

Table 5.1   Coding noise components and signs for all 20 test conditions

	1	2	3	4	5	6	7	8	9	10	11	12	13	14	15	16	17	18	19	20
Y	+			+	+		+	+	+	+	−			−	−		−	−	−	−
U		+		+		+	+	+	−	−		−			−	−	+	+	−	−
V			+	+	+	+	−	+	−					−	−	−	+	−	+	−

along specific directions in YUV space in this fashion (van den Branden Lambrecht and Farrell, 1996). Examples of the resulting distortions are shown in Figures 5.1(b) and 5.1(c).

5.1.2 Subjective Experiments

Psychophysical data was collected for two subjects (GEM and JEF) using a QUEST procedure (Watson and Pelli, 1983). In forced-choice experiments, the subjects were shown the original image together with two test images, one of which was the distorted image, and the other one the original. Subjects had to identify the distorted image, and the percentage of correct answers was recorded for varying noise intensities (van den Branden Lambrecht and Farrell, 1996). The responses for two test conditions are shown in Figure 5.2.

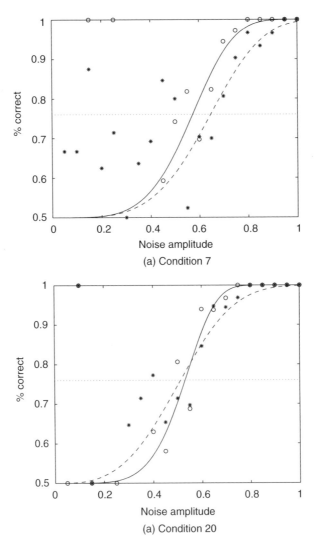

Figure 5.2 Percentage of correct answers versus noise amplitude and fitted psychometric functions for subjects GEM (stars, dashed curve) and JEF (circles, solid curve) for two test conditions. The dotted horizontal line indicates the detection threshold.

Such data can be modeled by the psychometric function

$$P(C) = 1 - 0.5 \, e^{-(x/\alpha)^{\beta}}, \tag{5.1}$$

where $P(C)$ is the probability of a correct answer, and x is the stimulus strength; α and β determine the midpoint and the slope of the function (Nachmias, 1981). These two parameters are estimated from the psychophysical data; the variable x represents the noise amplitude in this procedure. The resulting function can be used to map the noise amplitude onto the '% correct'-scale. Figure 5.2 also shows the results obtained in such a manner for two test conditions.

The detection threshold can now be determined from these data. Assuming an ideal observer model as discussed in section 4.2.6, the detection threshold can be defined as the observer detecting the distortion with a probability of 76%, which is virtually the same as the empirical 75%-threshold between chance and perfection in forced-choice experiments with two alternatives. This probability is indicated by the dotted horizontal line in Figure 5.2. The detection thresholds and their 95% confidence intervals for subjects GEM and JEF computed from the intersection of the estimated psychometric functions with the 76%-line for all 20 test conditions are shown in Figure 5.3. Even though some of the confidence intervals are quite large, the correlation between the thresholds of the two subjects is evident.

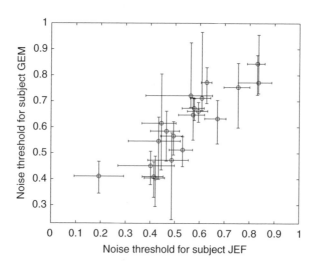

Figure 5.3 Detection thresholds of subject GEM versus subject JEF for all 20 test conditions. The error bars indicate the corresponding 95% confidence intervals.

5.1.3 Prediction Performance

For analyzing the performance of the perceptual distortion metric (PDM) from section 4.2 with respect to still images, the components of the metric pertaining to temporal aspects of vision, i.e. the temporal filters, are removed. Furthermore, the PDM has to be tuned to contrast sensitivity and masking data from psychophysical experiments with static stimuli.

Under certain assumptions for the ideal observer model (see section 4.2.6), the squared-error norm is equal to one at detection threshold, where the ideal observer is able to detect the distortion with a probability of 76% (Teo and Heeger, 1994a). The output of the PDM can thus be used to derive a threshold prediction by determining the noise amplitude at which the output of the metric is equal to its threshold value (this is not possible with PSNR, for example, as it does not have a predetermined value for the threshold of visibility). The scatter plot of PDM threshold predictions versus the estimated detection thresholds of the two subjects is shown in Figure 5.4. It can

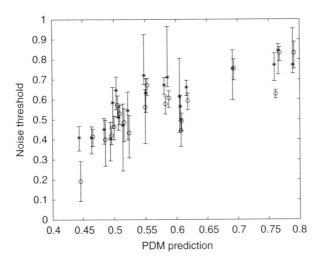

Figure 5.4 Detection thresholds of subjects GEM (stars) and JEF (circles) versus PDM predictions for all 20 test conditions. The error bars indicate the corresponding 95% confidence intervals.

be seen that the predictions of the metric are quite accurate for most of the test conditions. The RMSE between the threshold predictions of the PDM and the mean thresholds of the two subjects over all conditions is 0.07, compared to an inter-subject RMSE of 0.1, which underlines the differences between the two observers. The correlation between the PDM's threshold predictions and the average subjective thresholds is around 0.87, which is statistically equivalent to the inter-subject correlation. The threshold predictions are within the 95% confidence interval of at least one subject for nearly all test conditions. The remaining discrepancies can be explained by the fact that the subjective data for some test conditions are relatively noisy (the data shown in Figure 5.2 belong to the most reliable conditions), making it almost impossible in certain cases to compute a reliable estimate of the detection threshold. It should also be noted that while the range of distortions in this test was rather wide, only one test image was used. For these reasons, the still image evaluation presented in this section should only be regarded as a first validation of the metric. Our main interest is the application of the PDM to video, which is discussed in the remainder of this chapter.

5.2 VIDEO

5.2.1 Test Sequences

For evaluating the performance of the PDM with respect to video, experimental data collected within the framework of the Video Quality Experts Group (VQEG) is used. The PDM was one of the metrics submitted for evaluation to the first phase of tests (refer to section 3.5.3 for an overview of VQEG's program). The sequences used by VQEG and their characteristics are described here.

A set of 8-second scenes comprising both natural and computer-generated scenes with different characteristics (e.g. spatial detail, color, motion) was selected by independent labs. 10 scenes with a frame rate of 25 Hz and a resolution of 720×576 pixels as well as 10 scenes with a frame rate of 30 Hz and a resolution of 720×486 pixels were created in the format specified by ITU-R Rec. BT.601-5 (1995) for 4:2:2 component video. A sample frame of each scene is shown in Figures 5.5 and 5.6. The scenes were disclosed to the proponents only after the submission of their metrics.

The emphasis of the first phase of VQEG was out-of-service testing (meaning that the full uncompressed reference sequence is available to the metrics) of production- and distribution-class video. Accordingly, the test conditions listed in Table 5.2 comprise mainly MPEG-2 encoded sequences with different profiles, levels and other parameter variations, including encoder concatenation, conversions between analog and digital video, and transmission errors. In total, 20 scenes were encoded for 16 test conditions each.

Figure 5.5 VQEG 25-Hz test scenes.

Before the sequences were shown to subjective viewers or assessed by the metrics, a normalization was carried out on all test sequences in order to remove global temporal and spatial misalignments as well as global chroma and luma gains and offsets (VQEG, 2000). This was required by some of the metrics and could not be taken for granted because of the mixed analog and digital processing in certain test conditions.

5.2.2 Subjective Experiments

For the subjective experiments, VQEG adhered to ITU-R Rec. BT.500-11 (2002). Viewing conditions and setup, assessment procedures, and analysis

Figure 5.6 VQEG 30-Hz test scenes.

Table 5.2 VQEG test conditions

Number	Codec	Bitrate	Comments
1	Betacam	N/A	5 generations
2	MPEG-2	19-19-12 Mb/s	3 generations
3	MPEG-2	50 Mb/s	I-frames only, 7 generations
4	MPEG-2	19-19-12 Mb/s	3 generations with PAL/NTSC
5	MPEG-2	8-4.5 Mb/s	2 generations
6	MPEG-2	8 Mb/s	Composite PAL/NTSC
7	MPEG-2	6 Mb/s	
8	MPEG-2	4.5 Mb/s	Composite PAL/NTSC
9	MPEG-2	3 Mb/s	
10	MPEG-2	4.5 Mb/s	
11	MPEG-2	3 Mb/s	Transmission errors
12	MPEG-2	4.5 Mb/s	Transmission errors
13	MPEG-2	2 Mb/s	3/4 resolution
14	MPEG-2	2 Mb/s	3/4 horizontal resolution
15	H.263	768 kb/s	1/2 resolution
16	H.263	1.5 Mb/s	1/2 resolution

methods were drawn from this recommendation.[†] In particular, the Double Stimulus Continuous Quality Scale (DSCQS) (see section 3.3.3) was used for rating the sequences. The mean subjective rating differences between reference and distorted sequences, also known as differential mean opinion scores (DMOS), are used in the analyses that follow.

The subjective experiments were carried out in eight different laboratories. Four labs ran the tests with the 50-Hz sequences, and the other four with the 60-Hz sequences. Furthermore, each lab ran two separate tests for low-quality (conditions 8–16) and high-quality (conditions 1–9) sequences. The viewing distance was fixed at five times screen height. A total of 287 non-expert viewers participated in the experiments, and 25 830 individual ratings were recorded. Post-screening of the subjective data was performed in accordance with ITU-R Rec. BT.500-11 (2002) in order to discard unstable viewers.

The distribution of the mean rating differences and the corresponding 95% confidence intervals are shown in Figure 5.7. As can be seen, the quality range is not covered very uniformly; instead there is a heavy emphasis on low-distortion sequences (the median rating difference is 15). This has important implications for the performance of the metrics, which will be discussed below. The confidence intervals are very small (the median for the 95% confidence interval size is 3.6), which is due to the large number of viewers in the subjective tests and the strict adherence to the specified viewing conditions by each lab. For a more detailed discussion of the subjective experiments and their results, the reader is referred to the VQEG (2000) report.

5.2.3 Prediction Performance

The scatter plot of subjective DMOS versus PDM predictions is shown in Figure 5.8. It can be seen that the PDM is able to predict the subjective ratings well for most test cases. Several of its outliers belong to the lowest-bitrate (H.263) sequences of the test. As the metric is based on a threshold model of human vision, performance degradations for such clearly visible distortions can be expected. A number of other outliers are due to a single 50-Hz scene with a lot of movement. They are probably due to inaccuracies in the temporal filtering of the submitted version.

[†]See the VQEG subjective test plan at for details, http://www.vqeg.org/

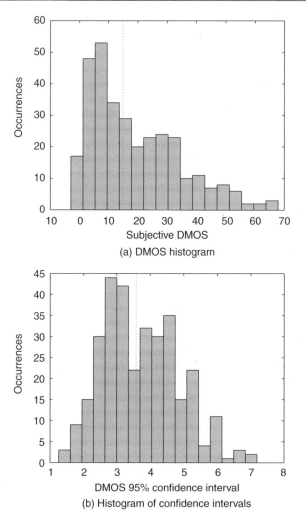

(a) DMOS histogram

(b) Histogram of confidence intervals

Figure 5.7 Distribution of differential mean opinion scores (a) and their 95% confidence intervals (b) over all test sequences. The dotted vertical lines denote the respective medians.

The DMOS-PDM plot should be compared with the scatter plot of DMOS versus PSNR in Figure 5.9. Because PSNR measures 'quality' instead of visual difference, the slope of the plot is negative. It can be observed that its spread is generally wider than for the PDM.

To put these plots in perspective, they have to be considered in relation to the reliability of subjective ratings. As discussed in section 3.3.2, perceived

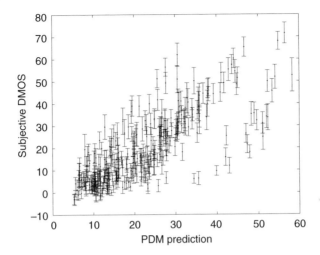

Figure 5.8 Perceived quality versus PDM predictions. The error bars indicate the 95% confidence intervals of the subjective ratings (from S. Winkler *et al.* (2001), Vision and video: Models and applications, in C. J. van den Branden Lambrecht (ed.), *Vision Models and Applications to Image and Video Processing*, chap. 10, Kluwer Academic Publishers. Copyright © 2001 Springer. Used with permission.).

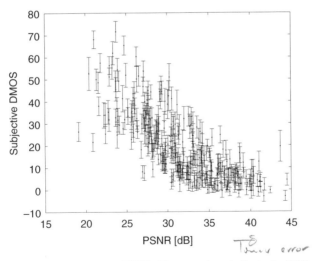

Figure 5.9 Perceived quality versus PSNR. The error bars indicate the 95% confidence intervals of the subjective ratings.

visual quality is an inherently subjective measure and can only be described statistically, i.e. by averaging over the opinions of a sufficiently large number of observers. Therefore the question is also how well subjects agree on the quality of a given image or video (this issue was also discussed in section 3.5.4).

As mentioned above, the subjective experiments for VQEG were carried out in eight different labs. This suggests taking a look at the agreement of ratings between different labs. An example of such an inter-lab DMOS scatter plot is shown in Figure 5.10. Although the confidence intervals are

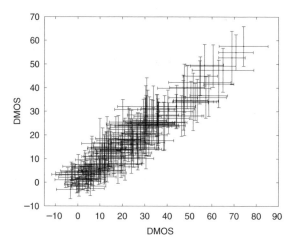

Figure 5.10 Example of inter-lab scatter plot of perceived quality. The error bars indicate the corresponding 95% confidence intervals.

larger due to the reduced number of subjects, there is a notable difference between it and Figures 5.8 and 5.9 in that the data points come to lie very close to a straight line.

These qualitative differences between the scatter plots can now be quantified with the help of the performance attributes described in section 3.5.1. Figure 5.11 shows the correlations between PDM predictions and subjective ratings over all sequences and for a number of subsets of test sequences, namely the 50-Hz and 60-Hz scenes, the low- and high-quality conditions as defined for the subjective experiments, the H.263 and non-H.263 sequences (conditions 15 and 16), the sequences with and without transmission errors (conditions 11 and 12), as well as the MPEG-only and non-MPEG sequences (conditions 2, 5, 7, 9, 10, 13, 14). As can be seen, the PDM can handle MPEG as well as non-MPEG kinds of distortions equally well and also behaves well with respect to sequences with transmission errors. Both the Pearson linear correlation and the Spearman rank-order correlation for most of the subsets are around 0.8. As mentioned before, the PDM performs worst for the H.263 sequences of the test.

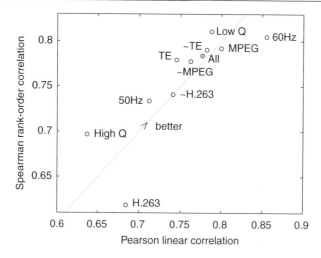

Figure 5.11 Correlations between PDM predictions and subjective ratings for several subsets of test sequences in the VQEG test, including all sequences, 50-Hz and 60-Hz scenes, low and high quality conditions, H.263 and non-H.263 sequences, sequences with and without transmission errors (TE), MPEG-only and non-MPEG sequences.

Comparisons of the PDM with the prediction performance of PSNR and the other metrics in the VQEG evaluation are given in Figure 5.12. Over all test sequences, there is not much difference between the top-performing metrics, which include the PDM, but also PSNR; in fact, their performance is statistically equivalent. Both Pearson and Spearman correlation are very close to 0.8 and go as high as 0.85 for certain subsets. The PDM does have one of the lowest outlier ratios for all subsets and is thus one of the most consistent metrics. The highest correlations are achieved by the PDM for the 60-Hz sequence set, for which the PDM outperforms all other metrics.

5.2.4 Discussion

Neither the PDM nor any of the other metrics were able to achieve the reliability of subjective ratings in the VQEG FR-TV Phase I evaluation. A surprise of this evaluation is probably the favorable prediction performance of PSNR with respect to other, much more complex metrics. A number of possible explanations can be given for this outcome. First, the range of distortions in the test is quite wide. Most metrics, however, had been designed for or tuned to a limited range (e.g. near threshold), so their prediction performance over all test conditions is reduced in relation to PSNR. Second, the data were collected for very specific viewing conditions.

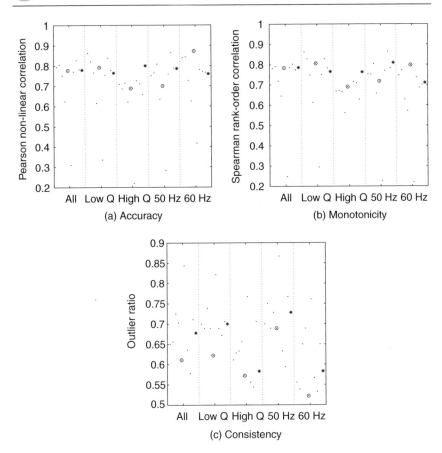

Figure 5.12 Comparison of the metrics in the VQEG evaluation with respect to three performance attributes (see section 3.5.1) for different subsets of sequences (optimal: high correlations, low outlier ratio). In every subset, each dot represents one of the ten participating metrics. The PDM is additionally marked with a circle, and PSNR is denoted with a star.

The PDM, for example, can adapt if these conditions are changed, whereas PSNR cannot. Third, PSNR is much more likely to fail in cases where distortions are not so 'benignly' and uniformly distributed among frames and color channels. Finally, the rigorous normalization of the test sequences with respect to alignment and luma/chroma gains or offsets may have given an additional advantage to PSNR. This will be investigated in depth in section 6.3 through different subjective experiments and test sequences.

While the Video Quality Experts Group needed to go through a second round of tests for successful standardization (see section 3.5.3), the value of

VQEG's first phase lies mainly in the creation of a framework for the reliable evaluation of video quality metrics. Furthermore, a large number of subjectively rated test sequences, which will also be used extensively in the remainder of this book, have been collected and made publicly available.[†]

5.3 COMPONENT ANALYSIS

5.3.1 Dissecting the PDM

The above-mentioned VQEG effort and other comparative studies have focused on evaluating the performance of entire video quality assessment systems. Hardly any analyses of single components of visual quality metrics have been published. Such an evaluation, which is important for achieving further improvements in this domain, is the purpose of this section. A number of implementation choices are analyzed that have to be made for most of today's quality assessment systems based on a vision model. These different implementations are equivalent from the point of view of simple threshold experiments, but can produce differing results for complex test sequences.

An example is the implementation of masking phenomena. Contrast gain control models such as the one used in the PDM (see section 4.2.4) have become quite popular in recent metrics. However, these models can be rather awkward to use in the general case, because they require a computation-intensive parameter fit for every change in the setup. Simpler models such as the so-called nonlinear transducer model[‡] are often more 'user-friendly', but are also less powerful. These and other models of spatial masking are discussed and compared by Klein *et al.* (1997) and Nadenau *et al.* (2002).

Another aspect of interest is the inclusion of contrast computation. Contrast is a relatively simple concept, but for complex stimuli a multitude of different mathematical contrast definitions have been proposed (see section 4.1.1). The importance of a local measure of contrast for natural images was shown in section 4.1, but which definition and which filter combination should be used to compute it?

Within the scope of this book, only a limited number of components can be investigated. Using the experimental data from the VQEG effort described above, the color space conversion stage, the perceptual decomposition, and

---

[†] See http://www.vqeg.org/

[‡] This three-parameter model divides the masking curve into a threshold range, where the target detection threshold is independent of masker contrast, and a masking range, where it grows with a certain power of the masker contrast.

the pooling and detection stage of the PDM (see Figure 4.6) are analyzed by comparing a number of different color spaces, decomposition filters, and some commonly used pooling algorithms in the following sections (Winkler, 2000). A similar evaluation of decomposition and pooling methods for an image quality metric was carried out recently by Fontaine *et al.* (2004).

5.3.2 Color Space

As discussed in section 4.2.2, the color processing in the PDM is based on an opponent color space proposed by Poirson and Wandell (1993, 1996). This particular color space was designed to separate color perception from pattern sensitivity, which has been considered an advantage for the modular design of the metric. However, it was derived from color-matching experiments and does not guarantee the perceptual uniformity of color differences, which is important for visual quality metrics. Color spaces such as CIE $L^*a^*b^*$ and CIE $L^*u^*v^*$ on the other hand (see Appendix for definitions), which have been used successfully in other metrics, were designed for color difference measurements, but lack pattern–color separability. Even simple YUV/YC_BC_R implements the opponent-color idea (Y encodes luminance, C_B the difference between the blue primary and luminance, and C_R the difference between the red primary and luminance) and provides the advantage of requiring no conversions from the digital component video input material (see, for example, Poynton (1996) for details about this color space), but it was not designed for measuring perceptual color differences.

The above-mentioned color spaces are similar in that they are all based on color differences. Therefore, they can be used interchangeably in the PDM by doing the respective color space conversion in the first module and ensuring that the threshold behavior of the metric does not change. In addition to evaluating the different color spaces, the full-color version of each implementation is also compared with its luminance-only version.

The results of this evaluation using the VQEG test sequences (see section 5.2.1) are shown in Figure 5.13. As can be seen, the differences in correlation are quite significant. Common to all color spaces is the fact that the additional consideration of the color components leads to a performance increase over the luminance-only version, although this improvement is not very large. In fact, the slight increases may not justify the double computational load imposed by the full-color PDM. However, one has to bear in mind that under most circumstances video encoders are 'good-natured' and distribute distortions more or less equally between the three color channels, therefore a result like this can be expected. Certain conditions with high

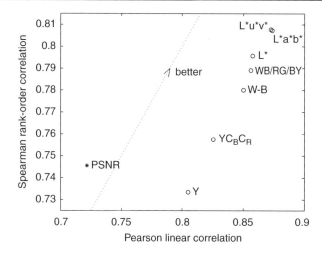

Figure 5.13 Correlations between PDM predictions and subjective ratings for different color spaces. PSNR is shown for comparison.

color saturation or unusually large distortions in the color channels may well be overlooked by a simple luminance metric, though.

Component video YC_BC_R exhibits the worst performance of the group. This is unfortunate, because it is the color space of the digital video input, so no further conversion is required. However, the conversions from YC_BC_R to the other color spaces incur only a relatively small penalty on the total computation time (on the order of a few percent) despite the nonlinearities involved. Furthermore, it is interesting to note that both CIE $L^*a^*b^*$ and CIE $L^*u^*v^*$ slightly outperform the Poirson–Wandell opponent color space (WB/RG/BY) in the PDM. This may be due to the better incorporation of perceived lightness and perceptual uniformity in these color spaces. The Poirson–Wandell opponent color space was chosen in the PDM because of its design for optimal pattern–color separability, which was supposed to facilitate the implementation of separate contrast sensitivity for each color channel. In the evaluation of natural video sequences, however, it turns out that this particular feature may only be of minor importance.

5.3.3 Decomposition Filters

Following the multi-channel theory of vision (see section 2.7), the PDM implements a decomposition of the input into a number of channels based on the spatio-temporal mechanisms in the visual system. As discussed in section 4.2.3, this perceptual decomposition is performed first in the temporal and then in the spatial domain.

First the temporal decomposition stage is investigated (see section 4.2.3). It was found that the specific filter types and lengths have no significant impact on prediction accuracy. Exchanging IIR filters with linear-phase FIR filters yields virtually identical PDM predictions. The approximation accuracy of the temporal mechanisms by the filters does not have a major influence, either. In fact, IIR filters with 2 poles and 2 zeros for the sustained mechanism and 4 poles and 4 zeros for the transient mechanism as well as FIR filters with 5 and 7 taps for the sustained and transient mechanism, respectively, leave the predictions of the PDM practically unchanged. This permits a further reduction of the delay of the PDM response. Finally, even the removal of the band-pass filter for the transient mechanism only reduces the correlations by a few percent.

The spatial decomposition in the PDM is taken care of by the steerable pyramid transform (see section 4.2.3). Many other filters have been proposed as approximations to the decomposition of visual information taking place in the human visual system, including Gabor filters (van den Branden Lambrecht and Verscheure, 1996), the Cortex transform (Daly, 1993), the DCT (Watson, 1998), and wavelets (Bolin and Meyer, 1999; Bradley, 1999; Lai and Kuo, 2000). We have found that the exact shape of the filters is not of paramount importance, but the goal here is also to obtain a good trade-off between implementation complexity, flexibility, and prediction accuracy. For use within a vision model, the steerable pyramid provides the advantage of rotation invariance, and it minimizes the amount of aliasing in the sub-bands. In the PDM, the basis filters have octave bandwidth and octave spacing; five sub-band levels with four orientation bands each plus one low-pass band are computed in each of the three color channels. Reduction or increase of the number of sub-band levels to four or six, respectively, does not lead to noticeable changes in the metric's prediction performance.

5.3.4 Pooling Algorithm

It is believed that the information represented in various channels of the primary visual cortex is integrated in higher-level areas of the brain. This process can be simulated by gathering the data from these channels according to rules of probability or vector summation, also known as pooling (Quick, 1974). However, little is known about the nature of the actual integration in the brain, and pooling mechanisms remain one of the most debated and uncertain aspects of vision modeling.

As discussed in section 4.2.5, mechanism responses can be combined by means of vector summation (also known as Minkowski summation or L_p-norm) using equation (4.29). Different exponents β in this equation have been found to yield good results for different experiments and implementations. $\beta = 2$ corresponds to the ideal observer formalism under independent Gaussian noise, which assumes that the observer has complete knowledge of the stimuli and uses a matched filter for detection (Teo and Heeger, 1994a). In a study of subjective experiments with coding artifacts, $\beta = 2$ was found to give good results (de Ridder, 1992). Intuitively, a few high distortions may draw the viewer's attention more than many lower ones. This behavior can be emphasized with higher exponents, which have been used in several other vision models, for example $\beta = 4$ (van den Branden Lambrecht, 1996b). The best fit of a contrast gain control model to masking data was achieved with $\beta = 5$ (Watson and Solomon, 1997).

In the PDM, pooling over channels and pixel locations is carried out with $\beta = 2$, whereas $\beta = 4$ is used for pooling over frames. We take a closer look at the latter part here. First, the temporal pooling exponent is varied between 0.1 and 6, and the correlations of PDM and subjective ratings are computed for the same set of sequences as in section 5.3.2. As can be seen from Figure 5.14(a), the maximum Pearson correlation $r_P = 0.857$ is obtained at $\beta = 2.9$, and the maximum Spearman correlation $r_S = 0.791$ at $\beta = 2.2$ (for comparison, the corresponding correlations for PSNR are $r_P = 0.72$ and $r_S = 0.74$). However, neither of the two peaks is very distinct. This result may be explained by the fact that the distortions are distributed quite uniformly over time for the majority of the test sequences, so that the individual predictions computed with $\beta = 0.1$ and $\beta = 6$ differ by less than 15%.

As an alternative, the distribution of ratings over frames can be used statistically to derive an overall rating. A simple method is to take the distortion rating that separates the lowest 80% of frame ratings from the highest 20%, for example. It can be argued that such a procedure emphasizes high distortions which are annoying to the viewer no matter how good the quality of the rest of the sequence is. Again, however, the specific histogram threshold chosen is rather arbitrary. Figure 5.14(b) shows the correlations computed for different values of this threshold. Here the influence is much more pronounced; the maximum Pearson correlation is obtained for thresholds between 55% and 75%, and the maximum Spearman correlation for thresholds between 45% and 65%, leading to the conclusion that a threshold of around 60% is the best choice overall for this method.

In any case, the pooling operation need not be carried out over all pixels in the entire sequence or frame. In order to take into account the focus of

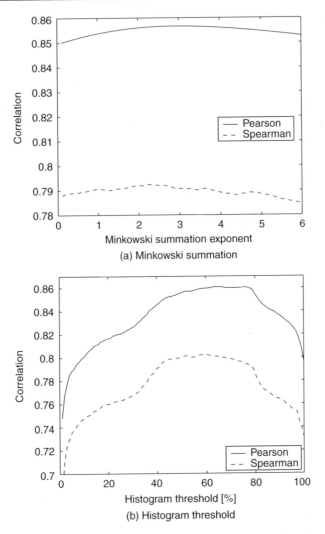

Figure 5.14 Pearson linear correlation (solid) and Spearman rank-order correlation (dashed) versus pooling exponent β (a) and versus histogram threshold (b).

attention of observers, for example, pooling can be carried out separately for spatio-temporal blocks of the sequence that cover roughly 100 milliseconds and two degrees of visual angle each (van den Branden Lambrecht and Verscheure, 1996). Alternatively, the distortion can be computed locally for every pixel, yielding perceptual distortion maps for better visualization of the temporal and spatial distribution of distortions, as demonstrated in

Figure 4.19. Such a distortion map can help the expert to locate and identify problems in the processing chain or shortcomings of an encoder, for example. This can be more useful and more reliable than a global measure in many quality assessment applications.

5.4 SUMMARY

The perceptual distortion metric (PDM) introduced in Chapter 4 was evaluated using still images and video sequences:

- First, the PDM has been validated using threshold data for color images, where its prediction performance is very close to the differences between subjects.
- With respect to video, the PDM has been shown to perform well over the wide range of scenes and test conditions from the VQEG evaluation. While its prediction performance is equivalent or even superior to other advanced video quality metrics, depending on the sequences considered, the PDM does not yet achieve the reliability of subjective ratings.
- The analysis of the different components of the PDM revealed that visual quality metrics which are essentially equivalent at the threshold level can exhibit significant differences in prediction performance for complex sequences, depending on the implementation choices made for the color space and the pooling algorithm used in the underlying vision model. The design of the decomposition filters on the other hand only has a negligible influence on the prediction accuracy.

In the following chapter, metric extensions will be discussed in an attempt to overcome the limitations of the PDM and other low-level vision-based distortion metrics and to improve their prediction performance.

6

Metric Extensions

The purpose of models is not to fit the data but to sharpen the questions.

Samuel Karlin

Several extensions of the PDM are explored in this chapter.

The first is the evaluation of blocking artifacts. The PDM is combined with an algorithm for blocking region segmentation to predict the perceived degree of blocking distortion. The prediction performance of the resulting perceptual blocking distortion metric (PBDM) is analyzed using data from subjective experiments on blockiness.

The second is the combination of the PDM with object segmentation. The necessary modifications of the metric are outlined, and the performance of the segmentation-supported PDM is evaluated using sequences on which face segmentation was performed.

Finally, the addition of attributes specifically related to visual quality instead of just visual fidelity are investigated. Sharpness and colorfulness are identified among these attributes and are quantified through the previously defined isotropic local contrast measure and the distribution of chroma in the sequence, respectively. The benefits of using these attributes are demonstrated with the help of additional test sequences and subjective experiments.

6.1 BLOCKING ARTIFACTS

6.1.1 Perceptual Blocking Distortion Metric

Some applications require more specific quality indicators than an overall rating or a visual distortion map. For instance, it can be useful to assess the

Digital Video Quality - Vision Models and Metrics   Stefan Winkler
© 2005 John Wiley & Sons, Ltd   ISBN: 0-470-02404-6

quality of certain image features such as contours, textures, blocking artifacts, or motion rendition (van den Branden Lambrecht, 1996b). Such specific quality ratings can be helpful in testing and fine-tuning encoders, for example. In particular, compression artifacts (see section 3.2.1) such as blockiness, ringing, or blur deserve a closer investigation. It is of interest to measure the perceived distortion caused by these different types of artifacts and to determine their influence on the overall quality degradation. Due to the popularity of the MPEG standard in digital video compression (see section 3.1.4), blocking artifacts are of particular importance. So far, however, metrics for blocking artifacts have focused mainly on still images (Miyahara and Kotani, 1985; Karunasekera and Kingsbury, 1995; Fränti, 1998).

Based on a modified version of the NVFM (Lindh and van den Branden Lambrecht, 1996) and the PDM (see section 4.2), a perceptual blocking distortion metric (PBDM) for digital video is proposed (Yu *et al.*, 2002). The underlying vision model has been simplified in that it works exclusively with luminance information (the chroma channels are disregarded), and the temporal part of the perceptual decomposition employs only one low-pass filter for the sustained mechanism (the transient mechanism is ignored). Furthermore, the mean value is subtracted from each channel after the temporal filtering. Another important difference is that no threshold data from psychophysical experiments are used to parameterize the model. Instead, the filter weights and contrast gain control parameters (see section 4.2.6) are chosen in a fitting process so as to maximize the Spearman rank-order correlation with part of the subjective data from the VQEG experiments (see section 5.2.2).

The PBDM relies on the fact that blocking artifacts, like other types of distortions, are dominant only in certain areas of a frame. These regions largely determine perceived blockiness. Therefore, the estimation of the distortion in these regions can serve as a measure of blocking artifacts. Based on this observation, the PBDM employs a segmentation stage to find regions where blocking artifacts dominate (see Figure 6.1).

Blocking region segmentation is carried out in the high-pass band of the steerable pyramid decomposition, where blocking artifacts are most pronounced. It consists of several steps (Yu *et al.*, 2002): First, horizontal and vertical edges are detected by looking for the specific pattern that block edges produce in the high-pass band. This edge detection is conducted both in the reference and the distorted sequence, and edges that exist in both are removed, because they must be due to the scene content. Likewise, edges shorter than 8 pixels are removed because of the DCT block size of

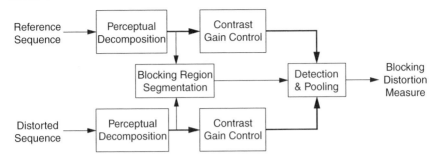

Figure 6.1 Block diagram of the perceptual blocking distortion metric (PBDM).

8×8 pixels in MPEG, as are immediately adjacent parallel edges. From this edge information, a blocking region map is created by extending the detected edges to the blocks most likely responsible for them. Finally, a ringing region map is created by looking for high-contrast edges in the reference sequence, which is then excluded from the blocking region map so that the final blocking region map represents only the areas in the sequence where blocking artifacts dominate. These segmentation steps make use of three thresholds, which are adjusted empirically such that the resulting blocking regions coincide with subjective assessment.

6.1.2 Test Sequences

Ten 60-Hz test scenes with a resolution of 720×486 pixels were selected from both the set described in ANSI-T1.801.01 (1995) and the VQEG test set (see section 5.2.1). The five ANSI scenes include *disgal* (a woman, mainly head and shoulders), *smity1* (a man in front of a more detailed background), *5row1* (a group of people at a table), *inspec* (a woman giving a presentation), and *ftball* (a high-motion football scene); they comprise 360 frames (12 seconds) each. The five VQEG scenes are the first five of Figure 5.6.

Each of the ANSI scenes was compressed with the MPEG-2 encoder of the MPEG Software Simulation Group (MSSG)[†] at bitrates of 768 kb/s, 1.4 Mb/s, 2 Mb/s and 3 Mb/s (the *ftball* scene was compressed at 5 Mb/s instead of 768 kb/s). For the VQEG scenes, the VQEG test conditions 9 (MPEG-2 at 3 Mb/s) and 14 (MPEG-2 at 2 Mb/s, 3/4 horizontal resolution) from Table 5.2 were used. This yielded a total of 30 test sequences.

[†]The source code is available at http://www.mpeg.org/home/~tristan/MPEG/MSSG/

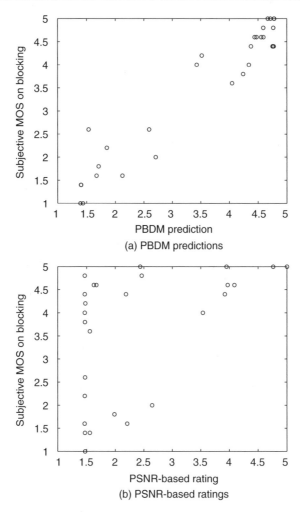

Figure 6.2 Perceived blocking impairment versus PBDM predictions (a) and PSNR-based ratings (b).

6.1.3 Subjective Experiments

Five subjects with normal or corrected-to-normal vision participated in the experiments (Yu *et al.*, 2002). They were asked to evaluate only the degree of blockiness in the sequence. Because of this specialized task, expert observers were chosen. Sequences were displayed on a 20-inch monitor, and the viewing distance was five times the display height.

The testing methodology adopted for the subjective experiments was variant II of the Double Stimulus Impairment Scale (DSIS-II) as defined in ITU-R Rec. BT.500-11 (2002). Its rating scale is the same as for the regular DSIS method, shown in Figure 3.8(b); the main difference is that the reference and the test sequence are repeated.

6.1.4 Prediction Performance

The scatter plot of perceived blocking distortion versus PBDM predictions is shown in Figure 6.2(a). The five-step DSIS rating scale was transformed to the numerical range from 1 (very annoying) to 5 (imperceptible) to compute the subjective mean opinion scores (MOS) on blocking, and the PBDM predictions Δ were transformed into the same range using the empirical formula $5 - \Delta^{0.6}$. As can be seen, there is a very good agreement between the metric's predictions and the subjective blocking ratings. The correlations are $r_P = 0.96$ and $r_S = 0.94$ (see section 3.5.1), which is as good as the agreement between different groups of observers discussed in section 5.2.3. It is also interesting to note that the commercial codecs used to create the VQEG test sequences are much better at minimizing blocking artifacts than the MSSG codec used for the ANSI sequences, but they produce noticeable blurring and ringing. The results show that the PBDM can successfully distinguish blocking artifacts from these other types of distortions.

For comparison, the scatter plot of perceived blocking distortion versus transformed PSNR-based ratings is shown in Figure 6.2(b). Here, the correlations are much worse, with $r_P = 0.49$ and $r_S = 0.51$. PSNR is thus unsuitable for measuring blocking artifacts, whereas the proposed perceptual blocking distortion metric can be considered a very reliable predictor of perceived blockiness.

6.2 OBJECT SEGMENTATION

While the previous sections were concerned mostly with lower-level aspects of vision, the cognitive behavior of people when watching video cannot be ignored in advanced quality metrics. However, cognitive behavior may differ greatly between individuals and situations, which makes it very difficult to generalize. Nevertheless, two important components should be pointed out, namely the shift of the focus of attention and the tracking of moving objects.

When watching video, we focus on particular areas of the scene. Studies have shown that the direction of gaze is not completely idiosyncratic to individual viewers. Instead, a significant number of viewers will focus on the same regions of a scene (Stelmach *et al.*, 1991; Stelmach and Tam, 1994; Endo *et al.*, 1994). Naturally, this focus of attention is highly scene-dependent. Maeder *et al.* (1996) as well as Osberger and Rohaly (2001) proposed constructing an importance map for the sequence as a prediction for the focus of attention, taking into account various perceptual factors such as edge strength, texture energy, contrast, color variation, homogeneity, etc.

In a similar manner, viewers may also track specific moving objects in a scene. In fact, motion tends to attract the viewers' attention. Now, the spatial acuity of the human visual system depends on the velocity of the image on the retina: as the retinal image velocity increases, spatial acuity decreases. The visual system addresses this problem by tracking moving objects with smooth-pursuit eye movements, which minimizes retinal image velocity and keeps the object of interest on the fovea. Smooth pursuit works well even for high velocities, but it is impeded by large accelerations and unpredictable motion (Eckert and Buchsbaum, 1993; Hearty, 1993). On the other hand, tracking a particular movement will reduce the spatial acuity for the background and objects moving in different directions or at different velocities. An appropriate adjustment of the spatio-temporal CSF as outlined in section 2.4.2 to account for some of these sensitivity changes can be considered as a first step in modeling such phenomena (Daly, 1998; Westen *et al.*, 1997).

Among the objects attracting most of our attention are people and especially human faces. If there are faces of people in a scene, we will look at them immediately. Furthermore, because of our familiarity with people's faces, we are very sensitive to distortions or artifacts occurring in them. The importance of faces is also underlined by a study of image appeal in consumer photography (Savakis *et al.*, 2000). People in the picture and their facial expressions are among the most important criteria for image selection. Furthermore, bringing out the structure and complexion of faces has been mentioned as an essential aspect of photography (Andrei, 1998, personal communication).

For these reasons, it makes sense to pay special attention to faces in visual quality assessment. Therefore, the combination of the PDM with face segmentation is explored. There exist relatively robust algorithms for face detection and segmentation (Gu and Bone, 1999), which are based on the fact that human skin colors are confined to a narrow region in the chrominance (C_B, C_R) plane, and their distribution is quite stable (Yang *et al.*, 1998). This greatly facilitates the detection of faces in images and sequences. It can then be followed by other object segmentation and tracking techniques to obtain reliable results across frames (Salembier and Marqués, 1999; Ziliani, 2000).

To take into account object segmentation with the PDM, a segmentation stage is added to find regions of interest, in this case faces. The output of the segmentation stage then guides the pooling process. The block diagram of the resulting segmentation-supported PDM is shown in Figure 6.3.

6.2.1 Test Sequences

Three test scenes shown in Figure 6.4 were selected. All contain faces at various scales and with various amounts of motion. Because of the small number of scenes, face segmentation was carried out by hand. For *fries* and *harp*, all 16 conditions from the VQEG experiments listed in Table 5.2 as well as the 8 conditions listed in Table 6.1 from the experiments described in section 6.3.4 were used. For *susie*, only the VQEG conditions were used, because this scene was not included in the other experiments. This yielded a total of 64 test sequences.

Table 6.1 Test conditions

Number	Codec	Version	Bitrate	Method
1	Intel Indeo Video	3.2	2 Mb/s	Vector quantization
2	Intel Indeo Video	4.5	2 Mb/s	Hybrid wavelet
3	Intel Indeo Video	5.11	1 Mb/s	Wavelet transform
4	Intel Indeo Video	5.11	2 Mb/s	Wavelet transform
5	MSSG MPEG-2	1.2	2 Mb/s	MC-DCT
6	Microsoft MPEG-4	2	1 Mb/s	MC-DCT
7	Microsoft MPEG-4	2	2 Mb/s	MC-DCT
8	Sorenson Video	2.11	2 Mb/s	Vector quantization

6.2.2 Prediction Performance

To evaluate the improvement of the prediction performance due to face segmentation, the ratings of the regular full-frame PDM are compared with those of the segmentation-supported PDM for the selection of test sequences described above in section 6.2.1. Using the regular PDM, the overall correlations for these sequences are $r_P = 0.82$ and $r_S = 0.79$ (see section 3.5.1).

When the segmentation of the sequences is added, the correlations rise to $r_P = 0.87$ and $r_S = 0.85$. The segmentation leads to a better agreement between the metric's predictions and the subjective ratings. As expected, the improvement is most noticeable for *susie*, in which the face covers a large part of the scene. Segmentation is least beneficial for *harp*, where the faces

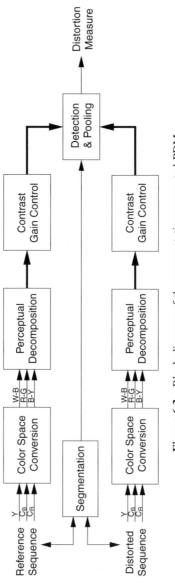

Figure 6.3 Block diagram of the segmentation-supported PDM.

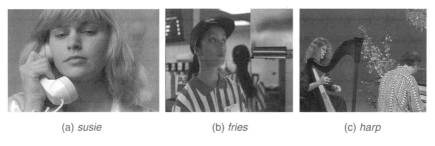

(a) *susie*   (b) *fries*   (c) *harp*

Figure 6.4   Segmentation test scenes.

are quite small and the strong distortions of the smooth background introduced by some test conditions are more annoying to viewers than in other
regions. Obviously, face segmentation alone is not sufficient for improving
the accuracy of PDM predictions in all cases, but the results show that it is
an important aspect.

6.3 IMAGE APPEAL

6.3.1 Background

As has become evident in Chapter 5, comparing a distorted sequence with its
original to derive a measure of quality has its limits with respect to prediction
accuracy, even if sophisticated and highly tuned models of the human visual
system are used. It was shown also in section 5.3 that further fine-tuning of
such metrics or their components for specific applications can improve the
prediction performance only slightly. Human observers, on the other hand,
seem to require no such 'tuning', yet are able to give much more reliable
quality ratings.

An important shortcoming of existing metrics is that they measure image
fidelity instead of perceived quality. This difference was discussed in section
3.3.2. The accuracy of the reproduction of the original on the display, even
considering the characteristics of the human visual system, is not the only
indicator of quality.

In an attempt to overcome the limitations that have been reached by
fidelity metrics, we therefore turn to more subjective attributes of image
quality, which we refer to as *image appeal* for better distinction. In a study of
image appeal in consumer photography, Savakis *et al.* (2000) compiled a list
of positive and negative influences in the ranking of pictures based on
experiments with human observers. Their results show that the most

important attributes for image selection are related to scene composition and location as well as the people in the picture and their expressions. Due to the high semantic level of these attributes, it is an extremely difficult and delicate task to take them into account with a general metric, however (see section 6.2).

Fortunately, a number of attributes that greatly influence the subjects' ranking decisions can be measured physically. In particular, colorful, well-lit, sharp pictures with high contrasts are considered attractive, whereas low-quality, dark and blurry pictures with low contrasts are often rejected (Savakis *et al.*, 2000). The depth of field, i.e. the separation between subject and background, and the range of colors and shades have also been mentioned as contributing factors (Chiossone, 1998, personal communication). The importance of high contrast and sharpness as well as colorfulness and saturation for good pictures has been confirmed by studies on naturalness (de Ridder *et al.*, 1995; Yendrikhovskij *et al.*, 1998) and has also been emphasized by professional photographers (Andrei, 1998, personal communication; Marchand, 1999, personal communication).

6.3.2 Quantifying Image Appeal

Based on the above-mentioned studies, *sharpness* and *colorfulness* are among the subjective attributes with the most significant influence on perceived quality. In order to work with these attributes, it is necessary to define them as measurable quantities.

6.3.2.1 Sharpness

For the computation of sharpness, we propose the use of a local contrast measure. The reasoning is that sharp images exhibit high contrasts, whereas blurring leads to a decrease in contrast. We employ the isotropic local contrast measure from section 4.1, which is based on the combination of analytic oriented filter responses. Because of its design properties, it is a natural measure of contrast in complex images.

For the computation of the isotropic local contrast according to equation (4.11), the filters described in section 4.1.4 are used. The remaining parameter is the level of the pyramidal decomposition. The lowest level is chosen here, because it contains the high-frequency information, which intuitively appears most suitable for the representation of sharpness. An example of the resulting isotropic local contrast is shown in Figure 6.5(a).

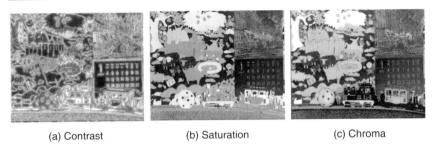

| (a) Contrast | (b) Saturation | (c) Chroma |

Figure 6.5 Luminance contrast C_0^l (a), saturation S_{uv} (b) and chroma C_{uv}^* (c) for a frame of the mobile scene (cf. Figure 6.7(a)).

To reduce the contrast values at every pixel of a sequence to a single number, pooling is carried out similar to the PDM (see section 4.2.5) by means of an L_p-norm. Several different exponents were tried, but best results were achieved with $p = 1$, i.e. plain averaging. Therefore, the sharpness rating of a sequence is defined as the mean isotropic local contrast over the entire sequence:

$$R_{sharp} = \mu_{C_0^l}. \tag{6.1}$$

6.3.2.2 Colorfulness

Colorfulness depends on two factors (Fedorovskaya *et al.*, 1997): the first factor is the average distance of image colors from a neutral gray, which may be modeled as the average chroma. The second factor is the distance between individual colors in the image, which may be modeled as the spread of the distribution of chroma values. If lightness differences between images are neglected, chroma can be replaced by saturation.

Conceptually, both saturation and chroma describe the purity of colors. *Saturation* is the colorfulness of an area judged in relation to its own brightness, and *chroma* is the colorfulness of an area judged in relation to the brightness of a similarly illuminated white area (Hunt, 1995). CIE $L^*u^*v^*$ color space (see Appendix) permits the computation of both measures. Saturation is defined using the u' and v' components from equation (4.3):

$$S_{uv} = 13\sqrt{(u' - u_0')^2 + (v' - v_0')^2}, \tag{6.2}$$

and chroma is defined as:

$$C_{uv}^* = \sqrt{u^{*2} + v^{*2}} = S_{uv}L^*. \tag{6.3}$$

These quantities are shown for a sample frame in Figures 6.5(b) and 6.5(c).

Several other color spaces with a saturation component exist. Examples are *HSI* (hue, saturation, intensity) (Gonzalez and Woods, 1992), *HSV* (hue, saturation, value) and *HLS* (hue, lightness, saturation) (Foley *et al.*, 1992). The saturation components in these color spaces are computed as follows:

$$S_{HSI} = 1 - \frac{3\min(R,G,B)}{R+G+B}, \tag{6.4}$$

$$S_{HSV} = \frac{\max(R,G,B) - \min(R,G,B)}{\max(R,G,B)}, \tag{6.5}$$

$$S_{HLS} = \begin{cases} \frac{\max(R,G,B)-\min(R,G,B)}{2L}, & \text{if } 0 \le L \le 0.5, \\ \frac{\max(R,G,B)-\min(R,G,B)}{2(1-L)}, & \text{if } 0.5 \le L \le 1, \end{cases} \tag{6.6}$$

where lightness $L = [\max(R,G,B) + \min(R,G,B)]/2$. The saturation of pure black is defined as $S = 0$ in all three color spaces, and $S = 1$ for pure colors red, green, blue, magenta, yellow, cyan.

S_{HSI}, S_{HSV}, and S_{HLS} are very similar and easy to compute. Chroma could also be defined as the product of saturation and lightness as in equation (6.3). However, these color spaces suffer from the fact that they are not perceptually uniform, and that they exhibit a singularity for black. Their saturation components were also used as a measure of colorfulness in the experiments described below, but the results obtained were generally better with saturation and chroma based on CIE $L^*u^*v^*$ color space from equations (6.2) and (6.3).

The best overall colorfulness ratings are obtained using the distribution of chroma values. This significantly reduces the number of outliers. According to the dependence of colorfulness on the chroma distribution parameters discussed above, the colorfulness rating of a sequence is thus defined as the sum of mean and standard deviation of chroma values over the entire sequence as suggested by Yendrikhovskij *et al.* (1998):

$$R_{\text{color}} = \mu_{C^*} + \sigma_{C^*}. \tag{6.7}$$

The underlying premise for using the sharpness and colorfulness ratings defined above as additional quality indicators is that a reduction of sharpness or colorfulness from the reference to the distorted sequence corresponds to a decrease in perceived quality. In other words, these differences $\Delta_{\text{sharp}} = R_{\text{sharp}} - \tilde{R}_{\text{sharp}}$ and $\Delta_{\text{color}} = R_{\text{color}} - \tilde{R}_{\text{color}}$ may be combined with the HVS-based distortion Δ_{PDM} for potentially more accurate predictions of overall visual quality. The benefits of such a combination will be investigated below.

A great advantage of these image appeal attributes is that they can be computed on the reference and the distorted sequences independently. This means that it is not necessary to have the entire reference sequence available at the testing site, but only its sharpness and colorfulness ratings, which can easily be transmitted together with the video data. They can thus be considered reduced-reference features.

6.3.3 Results with VQEG Data

The sharpness and colorfulness ratings were computed for the VQEG test sequences described in section 5.2.1. The results are compared with the overall subjective quality ratings from section 5.2.2 in Figure 6.6. As can be seen, there exists a correlation between the sharpness rating differences and the subjective quality ratings ($r_P = 0.63$, $r_S = 0.58$). The negative outliers are due almost exclusively to condition 1 (Betacam), which introduces noise and strong color artifacts, leading to an unusual increase of the sharpness rating.

Keep in mind that the sharpness rating was not conceived as an independent quality measure, but has to be combined with a fidelity metric such as the perceptual distortion metric (PDM) from section 4.2. This combination is implemented as $\Delta_{PDM} + w \max(0, \Delta_{sharp})$, so that negative differences are excluded, and the sharpness ratings are scaled to a range comparable to the PDM predictions. Using the optimum $w = 486$, the correlation with subjective quality ratings increases by 5% compared to PDM-only predictions (see final results in Figure 6.13). This shows that the additional consideration of sharpness by means of a contrast measure improves the prediction performance of the PDM.

The colorfulness rating differences, on the other hand, are negative for most sequences, which is counter-intuitive and seems to contradict the above-mentioned premise. Furthermore, they exhibit no correlation at all with subjective quality ratings (see Figure 6.6(b)), not even in combination with the PDM predictions. This can be explained by the rigorous normalization with respect to global chroma and luma gains and offsets that was carried out on the VQEG test sequences prior to the experiments (see section 5.2.1). When this normalization is reversed, the colorfulness rating differences become positive for most sequences, as expected. However, the normalization cannot be undone for the VQEG subjective ratings, which

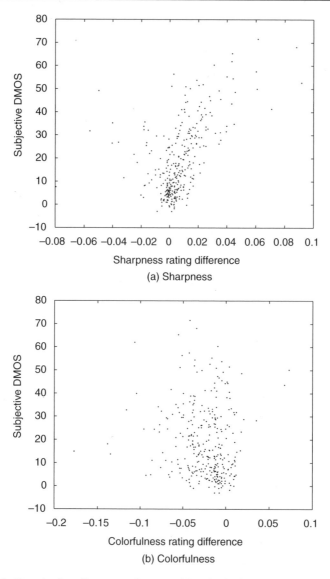

Figure 6.6 Perceived quality versus sharpness (a) and colorfulness (b) rating differences.

were collected using the normalized sequences. Therefore, no conclusion about the effectiveness of the colorfulness rating can be drawn from the VQEG data. Additional subjective experiments with unnormalized test sequences are necessary, which are described in the following.

6.3.4 Test Sequences

For evaluating the usefulness of sharpness and colorfulness ratings, subjective experiments were conducted with the test scenes shown in Figure 6.7 and the test conditions listed in Table 6.1.

(a) Scene 1: *mobile*   (b) Scene 2: *barcelona*   (c) Scene 3: *harp*

(d) Scene 4: *graphics*   (e) Scene 5: *canoe*   (f) Scene 6: *formula 1*

(g) Scene 7: *fries*   (h) Scene 8: *message*   (i) Scene 9: *rugby*

Figure 6.7   Test scenes.

The nine test scenes were selected from the set of VQEG scenes (see section 5.2.2) to include spatial detail, saturated colors, motion, and synthetic sequences. They are 8 seconds long with a frame rate of 25 Hz. They were de-interlaced and subsampled from the interlaced ITU-R Rec. BT.601-5 (2000) format to a resolution of 360×288 pixels per frame for progressive display. It should be noted that this led to slight aliasing artifacts in some of the scenes. Because of the DSCQS testing methodology used (see section 6.3.5), this should not affect the results of the experiment, however.

The codecs selected for creating the test sequences (see Table 6.1) are all implemented in software. Except for the MPEG-2 codec of the MPEG Software Simulation Group (MSSG),[†] they are DirectShow and QuickTime codecs. In contrast to the VQEG test conditions with a heavy focus on MPEG (see Table 5.2), these codecs use several different compression methods. Adobe Premiere[‡] was used for interfacing with the Windows codecs. A keyframe (I-frame) interval of 25 frames (1 second) was chosen. Two of the six codecs were operated at two different bitrates for comparison, yielding a total of eight test conditions and 72 test sequences. No normalization or calibration was carried out.

6.3.5 Subjective Experiments

The basis for the subjective experiments was again ITU-R Rec. BT.500-11 (2002). A total of 30 observers (23 males and 7 females) participated in the experiments. Their age ranged from 20 to 55 years; most of them were university students. The observers were tested for normal or corrected-to-normal vision with the help of a Snellen chart,[$] and for normal color vision using three Ishihara charts.[#]

A 19-inch ADI PD-959 MicroScan monitor was used for displaying the sequences. Its refresh rate was set to 85 Hz, and its screen resolution was set to 800×600 pixels, so that the sequences covered nearly one-quarter of the display area. A black level adjustment was carried out for a peak screen luminance of $70 \, \text{cd/m}^2$. The monitor gamma was determined through luminance measurements for different gray values y, which were approximated with the following function:

$$L(Y) = \alpha + \beta \left(\frac{Y}{255} \right)^{\gamma},$$

$$(6.8)$$

with $\alpha = -0.14 \, \text{cd/m}^2$, $\beta = 73.31 \, \text{cd/m}^2$, and $\gamma = 2.14$ (see Figure 6.8).

The Double Stimulus Continuous Quality Scale (DSCQS) method (see section 3.3.3) was selected for the experiments. The subjects were introduced to the method and their task, and training sequences were shown to demonstrate the range and type of impairments to be assessed.

[†]The source code is available at http://www.mpeg.org/home/~tristan/MPEG/MSSG/
[‡]See http://www.adobe.com/products/premiere/main.html for more information.
[$]Available at http://www.mdsupport.org/snellen.html
[#]Available at http://www.toledo-bend.com/colorblind/Ishihara.html

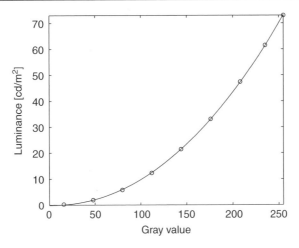

Figure 6.8 Screen luminance measurements (circles) and their approximation (curve).

The actual test sequences were presented to each observer in two sessions of 36 trials each. Their order was individually randomized so as to minimize effects of fatigue and adaptation. Windows Media Player 7[†] with a hand-written 'skin' (a uniform black background around the sequence) was used to display the sequences on the monitor. The viewing distance was 4–5 times the height of the active screen area.

After the experiments, post-screening of the subjective data was performed as specified in Annex 2 of ITU-R Rec. BT.500-11 (2002) to determine unstable viewers, but none of the subjects had to be removed.

The resulting differential mean opinion scores (DMOS) and their 95% confidence intervals for all 72 test sequences are shown in Figure 6.9. As can be seen, the entire quality range is covered quite uniformly (the median of the rating differences is 38), as was the intention of the test, and in contrast to the VQEG experiments (cf. Figure 5.7). The size of the confidence intervals is also satisfactory (median of 5.6). As a matter of fact, they are not much wider than in the VQEG experiments.

Figure 6.10 shows the subjective DMOS and confidence intervals, separated by scene and by condition. The separation by test scene reveals that scene 2 (*barcelona*) is the most critical one with the largest distortions averaged over conditions, followed by scenes 1 (*mobile*) and 3 (*harp*). Scenes 7 (*fries*) and 8 (*message*) on the other hand exhibit the smallest distortions.

[†]Available at http://www.microsoft.com/windows/windowsmedia/en/software/Playerv7.asp

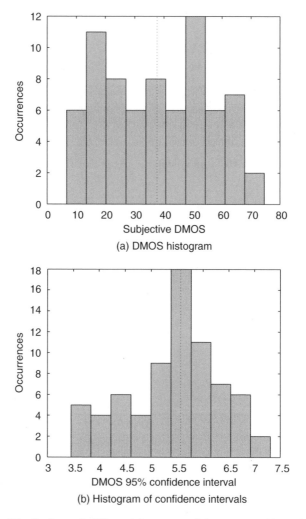

(a) DMOS histogram

(b) Histogram of confidence intervals

Figure 6.9 Distribution of differential mean opinion scores (a) and their 95% confidence intervals (b) over all test sequences. The dotted vertical lines denote the respective medians.

Several subjects mentioned that scene 8 (a horizontally scrolling message) actually was the most difficult test sequence to rate, and this is also where most confusions between reference and compressed sequence (i.e. negative rating differences) occurred.

It is instructive to compare the compression performance of the different codecs and their compression methods. The separation by test condition in Figure 6.10(b) shows that condition 5 (MPEG-2 at 2 Mb/s) exhibits the

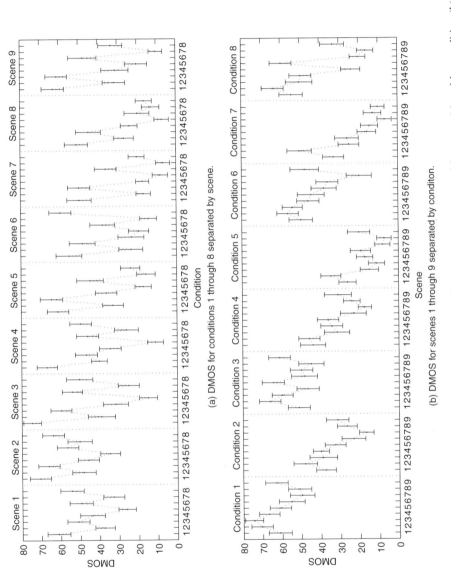

Figure 6.10 Subjective DMOS and confidence intervals for all test sequences separated by scene (a) and by condition (b).

highest quality over all scenes, closely followed by condition 7 (MPEG-4 at 2 Mb/s). At 1 Mb/s, the MPEG-4 codec (condition 6) outperforms conditions 1, 3, and 8. It should be noted that the Intel Indeo Video codecs and the Sorenson Video codec were designed for lower bitrates than the ones used in this test and obviously do not scale well at all, as opposed to MPEG-2 and MPEG-4. Comparing Figures 6.10(a) and 6.10(b) reveals that the perceived quality depends much more on the codec and bitrate than on the particular scene content in these experiments.

6.3.6 PDM Prediction Performance

Before returning to the image appeal attributes, let us take a look at the prediction performance of the regular PDM for these sequences. This is of interest for two reasons. First, as mentioned before, no normalization of the test sequences was carried out in this test. Second, the codecs and compression algorithms described above used to create the test sequences and the resulting visual quality of the sequences are very different from the VQEG test conditions (cf. Table 5.2). The latter rely almost exclusively on MPEG-2 and H.263, which are based on very similar compression algorithms (block-based DCT with motion compensation), whereas this test adds codecs based on vector quantization, the wavelet transform and hybrid methods. One of the advantages of the PDM is that it is independent of the compression method due to its underlying general vision model, contrary to specialized artifact metrics (cf. section 3.4.4).

The scatter plot of perceived quality versus PDM predictions is shown in Figure 6.11(a). It can be seen that the PDM is able to predict the subjective ratings well for most test sequences. The outliers belong mainly to conditions 1 and 8, the lowest-quality sequences in the test, as well as the computer-graphics scenes, where some of the Windows-based codecs introduced strong color distortions around the text, which was rated more severely by the subjects than by the PDM. It should be noted that performance degradations for such strong distortions can be expected, because the metric is based on a threshold model of human vision. Despite the much lower quality of the sequences compared to the VQEG experiments, the correlations between subjective DMOS and PDM predictions over all sequences are above 0.8 (see also final results in Figure 6.13).

The prediction performance of the PDM should be compared with PSNR, for which the corresponding scatter plot is shown in Figure 6.11(b). Because PSNR measures 'quality' instead of distortion, the slope of the plot is negative. It can be observed that its spread is wider than for the PDM, i.e.

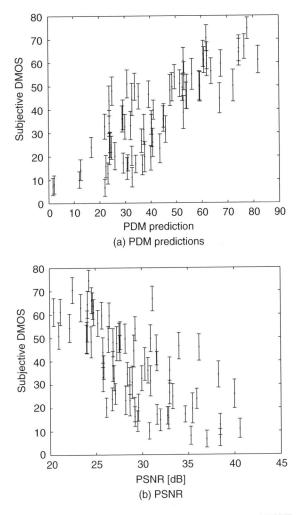

Figure 6.11   (a) Perceived quality versus PDM predictions (a) and PSNR (b). The error bars indicate the 95% confidence intervals of the subjective ratings.

there is a higher number of outliers. While PSNR achieved a performance comparable to the PDM in the VQEG test, its correlations have now decreased significantly to below 0.7.

6.3.7 Performance with Image Appeal Attributes

Now the benefits of combining the PDM quality predictions with the image appeal attributes are analyzed. The sharpness and colorfulness ratings are

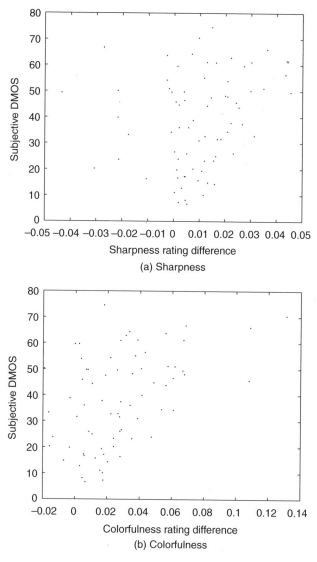

Figure 6.12 (a) Perceived quality versus sharpness (a) and colorfulness (b) rating differences.

computed for the test sequences described above in section 6.3.4. The results are compared with the subjective quality ratings from section 6.3.5 in Figure 6.12. The correlation between the subjective quality ratings and the sharpness rating differences is lower than for the VQEG sequences (see section 6.3.3). This is mainly due to the extreme outliers pertaining to conditions 1 and 8. These conditions introduce considerable distortions leading to additional strong edges in the compressed sequences, which increase the overall contrast.

On the other hand, a correlation between colorfulness rating differences and subjective quality ratings can now be observed. This confirms our assumption that the counter-intuitive behavior of the colorfulness ratings for the VQEG sequences was due to their rigorous normalization. Without such a normalization, the behavior is as expected for the test sequences described above in section 6.3.4, i.e. the colorfulness of the compressed sequences is reduced with respect to the reference for nearly all test sequences (see Figure 6.12(b)).

We stress again that neither the sharpness rating nor the colorfulness rating was designed as an independent measure of quality; both have to be used in combination with a visual fidelity metric. Therefore, the sharpness and colorfulness rating differences are combined with the output of the PDM as $\Delta_{PDM} + w_{sharp}\max(0, \Delta_{sharp}) + w_{color}\max(0, \Delta_{color})$. The rating differences are thus scaled to a range comparable to the PDM predictions, and negative differences are excluded. The results achieved with the optimum weights are shown in Figure 6.13.

It is evident that the additional consideration of sharpness and colorfulness improves the prediction performance of the PDM. The improvement with the sharpness rating alone is smaller than for the VQEG data. Together with the

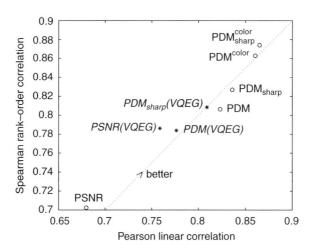

Figure 6.13 Prediction performance of the PDM alone and in combination with image appeal attributes for the VQEG test sequences (stars) as well as the new test sequences (circles). PSNR correlations are shown for comparison.

results discussed in section 6.3.3, this indicates that the sharpness rating is more useful for sequences with relatively low distortions. The colorfulness rating, on the other hand, which is of low computational complexity, gives a significant performance boost to the PDM predictions.

6.4 SUMMARY

A number of promising applications and extensions of the PDM were investigated in this chapter:

- A perceptual blocking distortion metric (PBDM) for evaluating the effects of blocking artifacts on perceived quality was described. Using a stage for blocking region segmentation, the PBDM was shown to achieve high correlations with subjective blockiness ratings.
- The usefulness of including object segmentation in the PDM was discussed. The advantages of segmentation support were demonstrated with test sequences showing human faces, resulting in better agreement of the PDM predictions with subjective ratings.
- Sharpness and colorfulness were identified as important attributes of image appeal. The attributes were quantified by defining a sharpness rating based on the measure of isotropic local contrast and a colorfulness rating derived from the distribution of chroma in the sequence. Extensive subjective experiments were carried out to establish a relationship between these ratings and perceived video quality. The results show that a combination of PDM predictions with the sharpness and colorfulness ratings leads to improvements in prediction performance.

7

Closing Remarks

We shall not cease from exploration
And the end of all our exploring
Will be to arrive where we started
And know the place for the first time.

T. S. Eliot

7.1 SUMMARY

Evaluating and optimizing the performance of digital imaging systems with respect to the capture, display, storage and transmission of visual information is one of the biggest challenges in the field of image and video processing. Understanding and modeling the characteristics of the human visual system is essential for this task.

We gave an overview of vision and discussed the anatomy and physiology of the human visual system in view of the applications investigated in this book. The following aspects can be emphasized: visual information is processed in different pathways and channels in the visual system, depending on its characteristics such as color, frequency, orientation, phase, etc. These channels play an important role in explaining interactions between stimuli. Furthermore, the response of the visual system depends much more on the contrast of patterns than on their absolute light levels. This makes the visual system highly adaptive. However, it is not equally sensitive to all stimuli.

We discussed the fundamentals of digital imaging systems. Image and video coding standards already exploit certain properties of the human visual

Digital Video Quality - Vision Models and Metrics   Stefan Winkler
© 2005 John Wiley & Sons, Ltd   ISBN: 0-470-02404-6

system to reduce bandwidth and storage requirements. Lossy compression as well as transmission errors lead to artifacts and distortions that affect video quality. Guaranteeing a certain level of quality has thus become an important concern for content providers. However, perceived quality depends on many different factors. It is inherently subjective and can only be described statistically.

We reviewed existing visual quality metrics. Pixel-based metrics such as MSE and PSNR are still popular despite their inability to give reliable predictions of perceived quality across different scenes and distortion types. Many vision-based quality metrics have been developed that provide a better prediction performance. However, independent comparison studies are rare, and so far no general-purpose metric has been found that is able to replace subjective testing.

Based on these foundations, we presented models of the human visual system and its characteristics in the framework of visual quality assessment and distortion minimization.

We constructed an isotropic local contrast measure by combining the responses of analytic directional filters. It is the first omnidirectional phase-independent contrast definition that can be applied to natural images and agrees well with perceived contrast.

We then described a perceptual distortion metric (PDM) for color video. The PDM is based on a model of the human visual system that takes into account color perception, the multi-channel architecture of temporal and spatial mechanisms, spatio-temporal contrast sensitivity, pattern masking, and channel interactions. It was shown to accurately fit data from psychophysical experiments.

The PDM was evaluated by means of subjective experiments using natural images and video sequences. It was validated using threshold data for color images, where its prediction performance is close to the differences between subjects. With respect to video, the PDM was shown to perform well over a wide range of scenes and test conditions. Its prediction performance is on a par with or even superior to other advanced video quality metrics, depending on the sequences considered. However, the PDM does not yet achieve the reliability of subjective ratings.

The analysis of the different components of the PDM revealed that visual quality metrics that are essentially equivalent at the threshold level can exhibit differences in prediction performance for complex sequences, depending on the implementation choices made for the color space and the pooling algorithm. The design of the decomposition filters on the other hand only has a negligible influence on the prediction accuracy.

We also investigated a number of promising metric extensions in an attempt to overcome the limitations of the PDM and other vision-based quality metrics and to improve their prediction performance. A perceptual blocking distortion metric (PBDM) for evaluating the effects of blocking artifacts was described. The PBDM was shown to achieve high correlations with perceived blockiness. Furthermore, the usefulness of including object segmentation in the PDM was discussed. The advantages of segmentation support were demonstrated with test sequences showing human faces, resulting in better agreement of the PDM predictions with subjective ratings.

Finally, we identified attributes of image appeal that contribute to perceived quality. The attributes were quantified by defining a sharpness rating based on the measure of isotropic local contrast and a colorfulness rating derived from the distribution of chroma in the sequence. Additional subjective experiments were carried out to establish a relationship between these ratings and perceived video quality. The results show that combining the PDM predictions with sharpness and colorfulness ratings leads to improvements in prediction performance.

7.2 PERSPECTIVES

The tools and techniques that were introduced in this book are quite general and may prove useful in a variety of image and video processing applications. Only a small number could be investigated within the scope of this book, and numerous extensions and improvements can be envisaged.

In general, the development of computational HVS-models itself is still in its infancy, and many issues remain to be solved. Most importantly, more comparative analyses of different modeling approaches are necessary. The collaborative efforts of *Modelfest* (Carney *et al.*, 2000, 2002) or the Video Quality Experts Group (VQEG, 2000, 2003) represent important steps in the right direction. Even if the former concerns low-level vision and the latter entire video quality assessment systems, both share the idea of applying different models to the same set of carefully selected subjective data under the same conditions. Such analyses will help determine the most promising approaches.

There are several modifications of the vision model underlying the perceptual distortion metric that can be considered:

- The spatio-temporal CSF used in the PDM is based on stabilized measurements and does not take into account natural unconstrained eye movements. This could be remedied using motion-compensated CSF models as proposed by Westen *et al.* (1997) or Daly (1998). This way, natural drift, smooth pursuit and saccadic eye movements can be integrated in the CSF.

- The contrast gain control model of pattern masking has a lot of potential for considering additional effects, in particular with respect to channel interactions and color masking. The measurements and models presented by Chen *et al.* (2000a,b) may be a good starting point. Another example is temporal masking, which has not received much attention so far, and which can be taken into account by adding a time dependency to the pooling function. Pertinent data are available that may facilitate the fitting of the corresponding model parameters (Boynton and Foley, 1999; Foley and Chen, 1999). Watson *et al.* (2001) incorporated certain aspects of temporal noise sensitivity and temporal masking into a video quality metric.

- Contrast masking may not be the optimal solution. With complex stimuli as are found in natural scenes, the distortion can be more noise-like, and masking can become much larger (Eckstein *et al.*, 1997; Blackwell, 1998). Entropy masking has been proposed as a bridge between contrast masking and noise masking, when the distortion is deterministic but unfamiliar (Watson *et al.*, 1997), which may be a good model for quality assessment by inexperienced viewers. Several different models for spatial masking are discussed and compared by Klein *et al.* (1997) and Nadenau *et al.* (2002).

- Finally, pattern adaptation has a distinct temporal component to it and is not taken into account by existing metrics. Ross and Speed (1991) presented a single-mechanisms model that accounts for both pattern adaptation and masking effects of simple stimuli. More recently, Meese and Holmes (2002) introduced a hybrid model of gain control that can explain adaptation and masking in a multi-channel setting.

It is important to realize that incremental vision model improvements and further fine-tuning alone may not lead to quantum leaps in prediction performance. In fact, such elaborate vision models have significant drawbacks. As mentioned before, human visual perception is highly adaptive, but also very dependent on certain parameters such as color and intensity of ambient lighting, viewing distance, media resolution, and others. It is possible to design HVS-models that try to meticulously incorporate all of these parameters. The problem with this approach is that the model becomes tuned to very specific situations, which is generally not practical. Besides, fitting the large number of free parameters to the necessary data is computationally very expensive due to iterative procedures required by the high degree of nonlinearity in the model. However, when looking at the example in Figure 3.9, the quality differences remain, even if viewing parameters such as background light or viewing distance are changed. It is clear that one will no longer be able to distinguish them from three meters away, but exactly here lies an answer to the problem: it is necessary to make realistic assumptions about the typical viewing conditions, and to derive from them a good model parameterization, which can actually work for a wide variety of situations.

Another problem with building and calibrating vision models is that most psychophysical experiments described in the literature focus on simple test stimuli like Gabor patches or noise patterns. This can only be a makeshift solution for the modeling of more complex phenomena that occur when viewing natural images. More studies, especially on masking, need to be done with complex scenes and patterns (Watson *et al.*, 1997; Nadenau *et al.*, 2002; Winkler and Süsstrunk, 2004).

Similarly, many psychophysical experiments have been carried out at threshold levels of vision, i.e. determining whether or not a certain stimulus is visible, whereas quality metrics and compression are often applied above threshold. This obvious discrepancy has to be overcome with supra-threshold experiments, otherwise the metrics run the risk of being nothing else than extrapolation guesses. Great care must be taken when using quality metrics based on threshold models and threshold data from simple stimuli for evaluating images or video with supra-threshold distortions. In fact, it may turn out that quality assessment of highly distorted video requires a completely new measurement paradigm.

This possible paradigm shift may actually be advantageous from the point of view of computational complexity. Like other HVS-based quality metrics, the proposed perceptual distortion metric is quite complex and requires a lot of computing power due to the extensive filtering and nonlinear operations in the underlying HVS-model. Dedicated hardware implementations can alleviate this problem to a certain extent, but such solutions are big and expensive and cannot be easily integrated into the average user's TV or mobile phone. Therefore, quality metrics may focus on specialized tasks or video material instead, for example specific codecs or artifacts, in order to keep complexity low while at the same time maintaining a good prediction performance. Several such metrics have been developed for blockiness (Winkler *et al.*, 2001; Wang *et al.*, 2002), blur (Marziliano *et al.*, 2004), and ringing (Yu *et al.*, 2000), for example.

Another important restriction of the PDM and other HVS-model based fidelity metrics is the need for the full reference sequence. In many applications the reference sequence simply cannot be made available at the testing site, for example somewhere out in the network, or a reference as such may not even exist, for instance at the output of the capture chip of a camera. Metrics are needed that rely only on a very limited amount of information about the reference, which can be transmitted along with the compressed bitstream, or even none at all. These reduced-reference or no-reference metrics would be much more versatile than full-reference metrics from an application point of view. However, they are less general than vision model-based metrics in the sense that they have to rely on certain assumptions about the sources and types of artifacts in order to make the quality predictions. This is the reason reduced-reference metrics (Wolf and Pinson, 1999; Horita *et al.*, 2003) and especially no-reference metrics (Coudoux *et al.*, 2001; Gastaldo *et al.*, 2002; Caviedes and Oberti, 2003; Winkler and Campos, 2003; Winkler and Dufaux, 2003) are usually based on the analysis of certain predefined artifacts or video features, which can then be related to overall quality for a specific application. The Video Quality Experts Group has already initiated evaluations of such reduced- and no-reference quality metrics.

Finally, vision may be the most essential of our senses, but it is certainly not the only one: we rarely watch video without sound. Focusing on visual quality alone cannot solve the problem of evaluating a multimedia experience, and the complex interactions between audio and video quality have been pointed out previously. Therefore, comprehensive audio-visual quality metrics are required that analyze both video and audio as well as their interactions. Only little work has been done in this area; the metrics described by Hollier and Voelcker (1997) or Jones and Atkinson (1998) are among the few examples in the literature to date.

As this concluding discussion shows, the future tasks in this area of research are challenging and need to be solved in close collaboration of experts in psychophysics, vision science and image processing.

Appendix: Color Space Conversions

Conversion from CIE 1931 XYZ tristimulus values to CIE $L^*a^*b^*$ and CIE $L^*u^*v^*$ color spaces is defined as follows (Wyszecki and Stiles, 1982). The conversions make use of the function

$$g(x) = \begin{cases} x^{1/3} & \text{if } x > 0.008856, \\ 7.787x + \frac{16}{116} & \text{otherwise.} \end{cases} \tag{A.1}$$

Both CIE $L^*a^*b^*$ and CIE $L^*u^*v^*$ space share a common lightness component L^*:

$$L^* = 116g(Y/Y_0) - 16. \tag{A.2}$$

The 0-subscript refers to the corresponding unit for the reference white being used. By definition, $L^* = 100$, $u^* = v^* = 0$, and $a^* = b^* = 0$ for the reference white.

The two chromaticity coordinates u^* and v^* in CIE $L^*u^*v^*$ space are computed as follows:

$$u^* = 13L^*(u' - u'_0), \ u' = \frac{4X}{X + 15Y + 3Z},$$
$$v^* = 13L^*(v' - v'_0), \ v' = \frac{9Y}{X + 15Y + 3Z}, \tag{A.3}$$

and the CIE $L^*u^*v^*$ color difference is given by

$$\Delta E^*_{uv} = \sqrt{(\Delta L^*)^2 + (\Delta u^*)^2 + (\Delta v^*)^2}. \tag{A.4}$$

Digital Video Quality - Vision Models and Metrics   Stefan Winkler
© 2005 John Wiley & Sons, Ltd   ISBN: 0-470-02404-6

The two chromaticity coordinates a^* and b^* in CIE $L^*a^*b^*$ space are computed as follows:

$$a^* = 500[g(X/X_0) - g(Y/Y_0)],$$
$$b^* = 200[g(Y/Y_0) - g(Z/Z_0)], \qquad \text{(A.5)}$$

and the CIE $L^*a^*b^*$ color difference is given by

$$\Delta E_{ab}^* = \sqrt{(\Delta L^*)^2 + (\Delta a^*)^2 + (\Delta b^*)^2}. \qquad \text{(A.6)}$$

References

All of the books in the world contain no more information than is broadcast as video in a single large American city in a single year. Not all bits have equal value.

Carl Sagan

Ahnelt, P. K. (1998). The photoreceptor mosaic. *Eye* **12**(3B):531–540.

Ahumada, A. J. Jr (1993). Computational image quality metrics: A review. In *SID Symposium Digest*, vol. 24, pp. 305–308.

Ahumada, A. J. Jr, Beard, B. L., Eriksson, R. (1998). Spatio-temporal discrimination model predicts temporal masking function. In *Proc. SPIE Human Vision and Electronic Imaging*, vol. 3299, pp. 120–127, San Jose, CA.

Ahumada, A. J. Jr, Null, C. H. (1993). Image quality: A multidimensional problem. In A. B. Watson (ed.), *Digital Images and Human Vision*, pp. 141–148, MIT Press.

Albrecht, D. G., Geisler, W. S. (1991). Motion selectivity and the contrast-response function of simple cells in the visual cortex. *Visual Neuroscience* **7**:531–546.

Aldridge, R. *et al.* (1995). Recency effect in the subjective assessment of digitally-coded television pictures. In *Proc. International Conference on Image Processing and its Applications*, pp. 336–339, Edinburgh, UK.

Alpert, T. (1996). The influence of the home viewing environment on the measurement of quality of service of digital TV broadcasting. In *MOSAIC Handbook*, pp. 159–163.

ANSI T1.801.01 (1995). Digital transport of video teleconferencing/video telephony signals – video test scenes for subjective and objective performance assessment. ANSI, Washington, DC.

Antoine, J.-P., Murenzi, R., Vandergheynst, P. (1999). Directional wavelets revisited: Cauchy wavelets and symmetry detection in patterns. *Applied and Computational Harmonic Analysis* **6**(3):314–345.

Ardito, M., Gunetti, M., Visca, M. (1996). Preferred viewing distance and display parameters. In *MOSAIC Handbook*, pp. 165–181.

Digital Video Quality - Vision Models and Metrics   Stefan Winkler
© 2005 John Wiley & Sons, Ltd   ISBN: 0-470-02404-6

Ascher, D., Grzywacz, N. M. (2000). A Bayesian model of temporal frequency masking. *Vision Research* **40**(16):2219–2232.

Avcibaş, İ., Sankur, B., Sayood, K. (2002). Statistical evaluation of image quality measures. *Journal of Electronic Imaging* **11**(2):206–223.

Bass, M. (ed. in chief) (1995). *Handbook of Optics: Fundamentals, Techniques, and Design*, 2nd edn, vol. 1, McGraw-Hill.

Baylor, D. A. (1987). Photoreceptor signals and vision. *Investigative Ophthalmology & Visual Science* **28**:34–49.

Beerends, J. G., de Caluwe, F. E. (1999). The influence of video quality on perceived audio quality and vice versa. *Journal of the Audio Engineering Society* **47**(5):355–362.

Blackwell, K. T. (1998). The effect of white and filtered noise on contrast detection thresholds. *Vision Research* **38**(2):267–280.

Blakemore, C. B., Campbell, F. W. (1969). On the existence of neurons in the human visual system selectively sensitive to the orientation and size of retinal images. *Journal of Physiology* **203**:237–260.

Bolin, M. R., Meyer, G. W. (1999). A visual difference metric for realistic image synthesis. In *Proc. SPIE Human Vision and Electronic Imaging*, vol. 3644, pp. 106–120, San Jose, CA.

Boynton, G. A., Foley, J. M. (1999). Temporal sensitivity of human luminance pattern mechanisms determined by masking with temporally modulated stimuli. *Vision Research* **39**(9):1641–1656.

Braddick, O., Campbell, F. W., Atkinson, J. (1978). Channels in vision: Basic aspects. In Held, R., Leibowitz, H. W., Teuber, H.-L. (eds), *Perception*, vol. 8 of *Handbook of Sensory Physiology*, pp. 3–38, Springer-Verlag.

Bradley, A. P. (1999). A wavelet visible difference predictor. *IEEE Transactions on Image Processing* **8**(5):717–730.

Brainard, D. H. (1995). Colorimetry. In Bass, M. (ed. in chief), *Handbook of Optics: Fundamentals, Techniques, and Design*, 2nd edn, vol. 1, chap. 26, McGraw-Hill.

Breitmeyer, B. G., Ogmen, H. (2000). Recent models and findings in visual backward masking: A comparison, review and update. *Perception & psychophysics* **72**(8):1572–1595.

Burbeck, C. A., Kelly, D. H. (1980). Spatiotemporal characteristics of visual mechanisms: Excitatory-inhibitory model. *Journal of the Optical Society of America* **70**(9):1121–1126.

Campbell, F. W., Gubisch, R. W. (1966). Optical quality of the human eye. *Journal of Physiology* **186**:558–578.

Campbell, F. W., Robson, J. G. (1968). Application of Fourier analysis to the visibility of gratings. *Journal of Physiology* **197**:551–566.

Carney, T., Klein, S. A., Hu, Q. (1996). Visual masking near spatiotemporal edges. In *Proc. SPIE Human Vision and Electronic Imaging*, vol. 2657, pp. 393–402, San Jose, CA.

Carney, T. *et al.* (2000). Modelfest: Year one results and plans for future years. In *Proc. SPIE Human Vision and Electronic Imaging*, vol. 3959, pp. 140–151, San Jose, CA.

Carney, T. *et al.* (2002). Extending the Modelfest image/threshold database into the spatio-temporal domain. In *Proc. SPIE Human Vision and Electronic Imaging*, vol. 4662, pp. 138–148, San Jose, CA.

Carpenter, R. H. S. (1988). *Movements of the Eyes*, Pion.

Caviedes, J. E., Oberti, F. (2003). No-reference quality metric for degraded and enhanced video. In *Proc. SPIE Visual Communications and Image Processing*, vol. 5150, pp. 621–632, Lugano, Switzerland.

Cermak, G. W. *et al.* (1998). Validating objective measures of MPEG video quality. *SMPTE Journal* **107**(4):226–235.

Charman, W. N. (1995). Optics of the eye. In Bass, M. (ed. in chief), *Handbook of Optics: Fundamentals, Techniques, and Design*, 2nd edn, vol. 1, chap. 24, McGraw-Hill.

Chen, C.-C., Foley, J. M., Brainard, D. H. (2000a). Detection of chromoluminance patterns on chromoluminance pedestals. I: Threshold measurements. *Vision Research* **40**(7): 773–788.

Chen, C.-C., Foley, J. M., Brainard, D. H. (2000b). Detection of chromoluminance patterns on chromoluminance pedestals. II: Model. *Vision Research* **40**(7):789–803.

Cole, G. R., Stromeyer III, C. F., Kronauer, R. E. (1990). Visual interactions with luminance and chromatic stimuli. *Journal of the Optical Society of America A* **7**(1):128–140.

Coudoux, F.-X., Gazalet, M. G., Derviaux, C., Corlay, P. (2001). Picture quality measurement based on block visibility in discrete cosine transform coded video sequences. *Journal of Electronic Imaging* **10**(2):498–510.

Curcio, C. A., Sloan, K. R., Kalina, R. E., Hendrickson, A. E. (1990). Human photoreceptor topography. *Journal of Comparative Neurology* **292**:497–523.

Curcio, C. A. *et al.* (1991). Distribution and morphology of human cone photoreceptors stained with anti-blue opsin. *Journal of Comparative Neurology* **312**:610–624.

Daly, S. (1993). The visible differences predictor: An algorithm for the assessment of image fidelity. In Watson, A. B. (ed.), *Digital Images and Human Vision*, pp. 179–206, MIT Press.

Daly, S. (1998). Engineering observations from spatiovelocity and spatiotemporal visual models. In *Proc. SPIE Human Vision and Electronic Imaging*, vol. 3299, pp. 180–191, San Jose, CA.

Daugman, J. G. (1980). Two-dimensional spectral analysis of cortical receptive field profiles. *Vision Research* **20**(10):847–856.

Daugman, J. G. (1985). Uncertainty relation for resolution in space, spatial frequency, and orientation optimized by two-dimensional visual cortical filters. *Journal of the Optical Society of America A* **2**(7):1160–1169.

Deffner, G. *et al.* (1994). Evaluation of display-image quality: Experts vs. non-experts. In *SID Symposium Digest*, vol. 25, pp. 475–478, Society for Information Display.

de Haan, G., Bellers, E. B. (1998). Deinterlacing – an overview. *Proceedings of the IEEE* **86**(9):1839–1857.

de Ridder, H. (1992). Minkowski-metrics as a combination rule for digital-image-coding impairments. In *Proc. SPIE Human Vision, Visual Processing and Digital Display*, vol. 1666, pp. 16–26, San Jose, CA.

de Ridder, H., Blommaert, F. J. J., Fedorovskaya, E. A. (1995). Naturalness and image quality: Chroma and hue variation in color images of natural scenes. In *Proc. SPIE Human Vision, Visual Processing and Digital Display*, vol. 2411, pp. 51–61, San Jose, CA.

De Valois, R. L., Smith, C. J., Kitai, S. T., Karoly, A. J. (1958). Electrical responses of primate visual system. I. Different layers of macaque lateral geniculate nucleus. *Journal of Comparative and Physiological Psychology.* **51**:662–668.

De Valois, R. L., Yund, E. W., Hepler, N. (1982a). The orientation and direction selecitivity of cells in macaque visual cortex. *Vision Research* 22(5):531–544.

De Valois, R. L., Albrecht, D. G., Thorell, L. G. (1982b). Spatial frequency selecitivity of cells in macaque visual cortex. *Vision Research* 22(5):545–559.

D'Zmura, M. *et al.* (1998). Contrast gain control for color image quality. In *Proc. SPIE Human Vision and Electronic Imaging*, vol. 3299, pp. 194–201, San Jose, CA.

EBU Broadcast Technology Management Committee (2002). The potential impact of flat panel displays on broadcast delivery of television. Technical Information I34, EBU, Geneva, Switzerland.

Eckert, M. P., Buchsbaum, G. (1993). The significance of eye movements and image acceleration for coding television image sequences. In Watson, A. B. (ed.), *Digital Images and Human Vision*, pp. 89–98, MIT Press.

Eckstein, M. P., Ahumada, A. J. Jr, Watson, A. B. (1997). Visual signal detection in structured backgrounds. II. Effects of contrast gain control, background variations, and white noise. *Journal of the Optical Society of America A* 14(9):2406–2419.

Endo, C., Asada, T., Haneishi, H., Miyake, Y. (1994). Analysis of the eye movements and its applications to image evaluation. In *Proc. Color Imaging Conference*, pp. 153–155, Scottsdale, AZ.

Engeldrum, P. G. (2000). *Psychometric Scaling: A Toolkit for Imaging Systems Development*, Imcotek Press.

Eriksson, R., Andrén, B., Brunnström, K. (1998). Modelling the perception of digital images: A performance study. In *Proc. SPIE Human Vision and Electronic Imaging*, vol. 3299, pp. 88–97, San Jose, CA.

Eskicioglu, A. M., Fisher, P. S. (1995). Image quality measures and their performance. *IEEE Transactions on Communications* 43(12):2959–2965.

Faugeras, O. D. (1979). Digital color image processing within the framework of a human visual model. *IEEE Transactions on Acoustics, Speech and Signal Processing* 27(4):380–393.

Fedorovskaya, E. A., de Ridder, H., Blommaert, F. J. J. (1997). Chroma variations and perceived quality of color images of natural scenes. *Color Research and Application* 22(2):96–110.

Field, D. J. (1987). Relations between the statistics of natural images and the response properties of cortical cells. *Journal of the Optical Society of America A* 4(12):2379–2394.

Foley, J. D., van Dam, A., Feiner, S. K., Hughes, J. F. (1992). *Computer Graphics. Principles and Practice*, 2nd edn, Addison-Wesley.

Foley, J. M. (1994). Human luminance pattern-vision mechanisms: Masking experiments require a new model. *Journal of the Optical Society of America A* 11(6):1710–1719.

Foley, J. M., Chen, C.-C. (1999). Pattern detection in the presence of maskers that differ in spatial phase and temporal offset: Threshold measurements and a model. *Vision Research* 39(23):3855–3872.

Foley, J. M., Yang, Y. (1991). Forward pattern masking: Effects of spatial frequency and contrast. *Journal of the Optical Society of America A* 8(12):2026–2037.

Fontaine, B., Saadane, H., Thomas, A. (2004). Perceptual quality metrics: Evaluation of individual components. In *Proc. International Conference on Image Processing*, pp. 3507–3510, Singapore.

Foster, K. H., Gaska, J. P., Nagler, M., Pollen, D. A. (1985). Spatial and temporal frequency selectivity of neurons in visual cortical areas V1 and V2 of the macaque monkey. *Journal of Physiology* **365**:331–363.

Fränti, P. (1998). Blockwise distortion measure for statistical and structural errors in digital images. *Signal Processing: Image Communication* **13**(2):89–98.

Fredericksen, R. E., Hess, R. F. (1997). Temporal detection in human vision: Dependence on stimulus energy. *Journal of the Optical Society of America A* **14**(10):2557–2569.

Fredericksen, R. E., Hess, R. F. (1998). Estimating multiple temporal mechanisms in human vision. *Vision Research* **38**(7):1023–1040.

Fuhrmann, D. R., Baro, J. A., Cox, J. R. Jr. (1995). Experimental evaluation of psychophysical distortion metrics for JPEG-coded images. *Journal of Electronic Imaging* **4**(4):397–406.

Gastaldo, P., Zunino, R., Rovetta, S. (2002). Objective assessment of MPEG-2 video quality. *Journal of Electronic Imaging* **11**(3):365–374.

Gescheider, G. A. (1997). *Psychophysics: The Fundamentals*, 3rd edn, Lawrence Erlbaum Associates.

Girod, B. (1989). The information theoretical significance of spatial and temporal masking in video signals. In *Proc. SPIE Human Vision, Visual Processing and Digital Display*, vol. 1077, pp. 178–187, Los Angeles, CA.

Gobbers, J.-F., Vandergheynst, P. (2002). Directional wavelet frames: Design and algorithms. *IEEE Transactions on Image Processing* **11**(4):363–372.

Gonzalez, R. C., Woods, R. E. (1992). *Digital Image Processing*, Addison-Wesley.

Graham, N., Sutter, A. (2000). Normalization: Contrast-gain control in simple (Fourier) and complex (non-Fourier) pathways of pattern vision. *Vision Research* **40**(20):2737–2761.

Grassmann, H. G. (1853). Zur Theorie der Farbenmischung. *Annalen der Physik und Chemie* **89**:69–84.

Green, D. M., Swets, J. A. (1966). *Signal Detection Theory and Psychophysics*, John Wiley.

Greenlee, M. W., Thomas, J. P. (1992). Effect of pattern adaptation on spatial frequency discrimination. *Journal of the Optical Society of America A* **9**(6):857–862.

Gu, L., Bone, D. (1999). Skin colour region detection in MPEG video sequences. In *Proc. International Conference on Image Analysis and Processing*, pp. 898–903, Venice, Italy.

Guyton, A. C. (1991). *Textbook of Medical Physiology*, 7th edn, W. B. Saunders.

Hammett, S. T., Smith, A. T. (1992). Two temporal channels or three? A reevaluation. *Vision Research* **32**(2):285–291.

Hearty, P. J. (1993). Achieving and confirming optimum image quality. In Watson, A. B. (ed.), *Digital Images and Human Vision*, pp. 149–162, MIT Press.

Hecht, E. (1997). *Optics*, 3rd edn, Addison-Wesley.

Hecht, S., Schlaer, S., Pirenne, M. H. (1942). Energy, quanta and vision. *Journal of General Physiology* **25**:819–840.

Heeger, D. J. (1992a). Half-squaring in responses of cat striate cells. *Visual Neuroscience* **9**:427–443.

Heeger, D. J. (1992b). Normalization of cell responses in cat striate cortex. *Visual Neuroscience* **9**:181–197.

Hering, E. (1878). *Zur Lehre vom Lichtsinne*, Carl Gerolds.

Hess, R. F., Snowden, R. J. (1992). Temporal properties of human visual filters: Number, shapes and spatial covariation. *Vision Research* **32**(1):47–59.

Hollier, M. P., Voelcker, R. (1997). Towards a multi-modal perceptual model. *BT Technology Journal* **15**(4):162–171.

Hood, D. C., Finkelstein, M. A. (1986). Sensitivity to light. In Boff, K. R., Kaufman, L., Thomas, J. P. (eds), *Handbook of Perception and Human Performance*, vol. 1, chap. 5, John Wiley.

Horita, Y. *et al.* (2003). Evaluation model considering static-temporal quality degradation and human memory for SSCQE video quality. In *Proc. SPIE Visual Communications and Image Processing*, vol. 5150, pp. 1601–1611, Lugano, Switzerland.

Hubel, D. H. (1995). *Eye, Brain, and Vision*, Scientific American Library.

Hubel, D. H., Wiesel, T. N. (1959). Receptive fields of single neurons in the cat's striate cortex. *Journal of Physiology* **148**:574–591.

Hubel, D. H., Wiesel, T. N. (1962). Receptive fields, binocular interaction and functional architecture in the cat's visual cortex. *Journal of Physiology* **160**:106–154.

Hubel, D. H., Wiesel, T. N. (1968). Receptive fields and functional architecture of monkey striate cortex. *Journal of Physiology* **195**:215–243.

Hubel, D. H., Wiesel, T. N. (1977). Functional architecture of macaque striate cortex. *Proceedings of the Royal Society of London B* **198**:1–59.

Hunt, R. W. G. (1995). *The Reproduction of Colour*, 5th edn, Fountain Press.

Hurvich, L. M., Jameson, D. (1957). An opponent-process theory of color vision. *Psychological Review* **64**:384–404.

ITU-R Recommendation BT.500-11 (2002). Methodology for the subjective assessment of the quality of television pictures. ITU, Geneva, Switzerland.

ITU-R Recommendation BT.601-5 (1995). Studio encoding parameters of digital television for standard 4:3 and wide-screen 16:9 aspect ratios. ITU, Geneva, Switzerland.

ITU-R Recommendation BT.709-5 (2002). Parameter values for the HDTV standards for production and international programme exchange. ITU, Geneva, Switzerland.

ITU-R Recommendation BT.1683 (2004). Objective perceptual video quality measurement techniques for standard definition digital broadcast television in the presence of a full reference. ITU, Geneva, Switzerland.

ITU-T Recommendation H.263 (1998). Video coding for low bit rate communication. ITU, Geneva, Switzerland.

ITU-T Recommendation H.264 (2003). Advanced video coding for generic audiovisual services. ITU, Geneva, Switzerland.

ITU-T Recommendation J.144 (2004). Objective perceptual video quality measurement techniques for digital cable television in the presence of a full reference. ITU, Geneva, Switzerland.

ITU-T Recommendation P.910 (1999). Subjective video quality assessment methods for multimedia applications. ITU, Geneva, Switzerland.

Jacobson, R. E., (1995). An evaluation of image quality metrics. *Journal of Photographic Science* **43**(1):7–16.

Jameson, D., Hurvich, L. M. (1955). Some quantitative aspects of an opponent-colors theory. I. Chromatic responses and spectral saturation. *Journal of the Optical Society of America* **45**(7):546–552.

Joly, A., Montard, N., Buttin, M. (2001). Audio-visual quality and interactions between television audio and video. In *Proc. International Symposium on Signal Processing and its Applications*, pp. 438–441, Kuala Lumpur, Malaysia.

Jones, C., Atkinson, D. J. (1998). Development of opinion-based audiovisual quality models for desktop video-teleconferencing. In *Proc. International Workshop on Quality of Service*, pp. 196–203, Napa Valley, CA.

Karunasekera, S. A., Kingsbury, N. G. (1995). A distortion measure for blocking artifacts in images based on human visual sensitivity. *IEEE Transactions on Image Processing* **4**(6):713–724.

Kelly, D. H. (1979a). Motion and vision. I. Stabilized images of stationary gratings. *Journal of the Optical Society of America* **69**(9):1266–1274.

Kelly, D. H. (1979b). Motion and vision. II. Stabilized spatio-temporal threshold surface. *Journal of the Optical Society of America* **69**(10):1340–1349.

Kelly, D. H. (1983). Spatiotemporal variation of chromatic and achromatic contrast thresholds. *Journal of the Optical Society of America* **73**(6):742–750.

Klein, S. A. (1993). Image quality and image compression: A psychophysicist's viewpoint. In Watson, A. B. (ed.), *Digital Images and Human Vision*, pp. 73–88, MIT Press.

Klein, S. A., Carney, T., Barghout-Stein, L., Tyler, C. W. (1997). Seven models of masking. In *Proc. SPIE Human Vision and Electronic Imaging*, vol. 3016, pp. 13–24, San Jose, CA.

Koenderink, J. J., van Doorn, A. J. (1979). Spatiotemporal contrast detection threshold surface is bimodal. *Optics Letters* **4**(1):32–34.

Kuffler, S. W. (1953). Discharge pattern and functional organisation of mammalian retina. *Journal of Neurophysiology* **16**:37–68.

Kutter, M., Winkler, S. (2002). A vision-based masking model for spread-spectrum image watermarking. *IEEE Transactions on Image Processing* **11**(1):16–25.

Lai, Y.-K., Kuo, C.-C. J. (2000). A Haar wavelet approach to compressed image quality measurement. *Visual Communication and Image Representation* **11**(1):17–40.

Lee, S., Pattichis, M. S., Bovik, A. C. (2002). Foveated video quality assessment. *IEEE Transactions on Multimedia* **4**(1):129–132.

Legge, G. E., Foley, J. M. (1980). Contrast masking in human vision. *Journal of the Optical Society of America* **70**(12):1458–1471.

Lehky, S. R. (1985). Temporal properties of visual channels measured by masking. *Journal of the Optical Society of America A* **2**(8):1260–1272.

Li, B., Meyer, G. W., Klassen, R. V. (1998). A comparison of two image quality models. In *Proc. SPIE Human Vision and Electronic Imaging*, vol. 3299, pp. 98–109, San Jose, CA.

Liang, J., Westheimer, G. (1995). Optical performances of human eyes derived from double-pass measurements. *Journal of the Optical Society of America A* **12**(7):1411–1416.

Lindh, P., van den Branden Lambrecht, C. J. (1996). Efficient spatio-temporal decomposition for perceptual processing of video sequences. In *Proc. International Conference on Image Processing*, vol. 3, pp. 331–334, Lausanne, Switzerland.

Lodge, N. (1996). An introduction to advanced subjective assessment methods and the work of the MOSAIC consortium. In *MOSAIC Handbook*, pp. 63–78.

Losada, M. A., Mullen, K. T. (1994). The spatial tuning of chromatic mechanisms identified by simultaneous masking. *Vision Research* **34**(3):331–341.

Losada, M. A., Mullen, K. T. (1995). Color and luminance spatial tuning estimated by noise masking in the absence of off-frequency looking. *Journal of the Optical Society of America A* **12**(2):250–260.

Lu, Z. *et al.* (2003). PQSM-based RR and NR video quality metrics. In *Proc. SPIE Visual Communications and Image Processing*, vol. 5150, pp. 633–640, Lugano, Switzerland.

Lubin, J. (1995). A visual discrimination model for imaging system design and evaluation. In Peli, E. (ed.), *Vision Models for Target Detection and Recognition*, pp. 245–283, World Scientific Publishing.

Lubin, J., Fibush, D. (1997). Sarnoff JND vision model. T1A1.5 Working Group Document #97-612, ANSI T1 Standards Committee.

Lukas, F. X. J., Budrikis, Z. L. (1982). Picture quality prediction based on a visual model. *IEEE Transactions on Communications* **30**(7):1679–1692.

Lund, A. M. (1993). The influence of video image size and resolution on viewing-distance preferences. *SMPTE Journal* **102**(5):407–415.

Maeder, A., Diederich, J., Niebur, E. (1996). Limiting human perception for image sequences. In *Proc. SPIE Human Vision and Electronic Imaging*, vol. 2657, pp. 330–337, San Jose, CA.

Mallat, S. (1998). *A Wavelet Tour of Signal Processing*. Academic Press.

Mallat, S., Zhong, S. (1992). Characterization of signals from multiscale edges. *IEEE Transactions on Pattern Analysis and Machine Intelligence* **14**(7):710–732.

Malo, J., Pons, A. M., Artigas, J. M. (1997). Subjective image fidelity metric based on bit allocation of the human visual system in the DCT domain. *Image and Vision Computing*, **15**(7):535–548.

Mandler, M. B., Makous, W. (1984). A three-channel model of temporal frequency perception. *Vision Research* **24**(12):1881–1887.

Mannos, J. L., Sakrison, D. J. (1974). The effects of a visual fidelity criterion on the encoding of images. *IEEE Transactions on Information Theory* **20**(4):525–536.

Marimont, D. H., Wandell, B. A. (1994). Matching color images: The effects of axial chromatic aberration. *Journal of the Optical Society of America A* **11**(12): 3113–3122.

Marmolin, H. (1986). Subjective MSE measures. *IEEE Transactions on Systems, Man, and Cybernetics* **16**(3):486–489.

Martens, J.-B., Meesters, L. (1998). Image dissimilarity. *Signal Processing* **70**(3):155–176.

Marziliano, P., Dufaux, F., Winkler, S., Ebrahimi, T. (2004). Perceptual blur and ringing metrics: Application to JPEG2000. *Signal Processing: Image Communication* **19**(2):163–172.

Masry, M. A., Hemami, S. S. (2004). A metric for continuous quality evaluation of compressed video with severe distortions. *Signal Processing: Image Communication* **19**(2):133–146.

Mayache, A., Eude, T., Cherifi, H. (1998). A comparison of image quality models and metrics based on human visual sensitivity. In *Proc. International Conference on Image Processing*, vol. 3, pp. 409–413, Chicago, IL.

Meese, T. S., Holmes, D. J. (2002). Adaptation and gain pool summation: Alternative models and masking data. *Vision Research* **42**(9):1113–1125.

Meese, T. S., Williams, C. B. (2000). Probability summation for multiple patches of luminance modulation. *Vision Research* **40**(16):2101–2113.

Michelson, A. A. (1927). *Studies in Optics*, University of Chicago Press.

Miyahara, M., Kotani, K. (1985). Block distortion in orthogonal transform coding – analysis, minimization and distortion measure. *IEEE Transactions on Communications* **33**(1):90–96.

Miyahara, M., Kotani, K., Algazi, V. R. (1998). Objective picture quality scale (PQS) for image coding. *IEEE Transactions on Communications* **46**(9):1215–1226.

MOSAIC (1996). A new single stimulus quality assessment methodology. RACE R2111.

Mullen, K. T. (1985). The contrast sensitivity of human colour vision to red-green and blue-yellow chromatic gratings. *Journal of Physiology* **359**:381–400.

Muschietti, M. A., Torrésani, B. (1995). Pyramidal algorithms for Littlewood Paley decompositions. *SIAM Journal of Mathematical Analysis* **26**(4):925–943.

Nachmias, J. (1981). On the psychometric function for contrast detection. *Vision Research* **21**:215–223.

Nadenau, M. J., Reichel, J., Kunt, M. (2002). Performance comparison of masking models based on a new psychovisual test method with natural scenery stimuli. *Signal Processing: Image Communication* **17**(10):807–823.

Olzak, L. A., Thomas, J. P. (1986). Seeing spatial patterns. In Boff, K. R., Kaufman, L., Thomas, J. P. (eds), *Handbook of Perception and Human Performance*, vol. 1, chap. 7, John Wiley.

Osberger, W., Rohaly, A. M. (2001). Automatic detection of regions of interest in complex video sequences. In *Proc. SPIE Human Vision and Electronic Imaging*, vol. 4299, pp. 361–372, San Jose, CA.

Peli, E. (1990). Contrast in complex images. *Journal of the Optical Society of America A* **7**(10):2032–2040.

Peli, E. (1997). In search of a contrast metric: Matching the perceived contrast of Gabor patches at different phases and bandwidths. *Vision Research* **37**(23):3217–3224.

Pelli, D. G., Farell, B. (1995). Psychophysical methods. In Bass, M. (ed. in chief), *et al. Handbook of Optics: Fundamentals, Techniques, and Design*, 2nd edn, vol. 1, chap. 29, McGraw-Hill.

Phillips, G. C., Wilson, H. R. (1984). Orientation bandwidth of spatial mechanisms measured by masking. *Journal of the Optical Society of America A* **1**(2):226–232.

Pinson, M. H., Wolf, S. (2004). The impact of monitor resolution and type on subjective video quality testing. NTIA Technical Memorandum TM-04-412, NTIA/ITS.

Poirson, A. B., Wandell, B. A. (1993). Appearance of colored patterns: Pattern-color separability. *Journal of the Optical Society of America A* **10**(12):2458–2470.

Poirson, A. B., Wandell, B. A. (1996). Pattern-color separable pathways predict sensitivity to simple colored patterns. *Vision Research* **36**(4):515–526.

Poynton, C. A. (1996). *A Technical Introduction to Digital Video*, John Wiley.

Poynton, C. (1998). The rehabilitation of gamma. In *Proc. SPIE Human Vision and Electronic Imaging*, vol. 3299, pp. 232–249, San Jose, CA.

Quick, R. R. Jr (1974). A vector-magnitude model of contrast detection. *Kybernetik* **16**:65–67.

Rihs, S. (1996). The influence of audio on perceived picture quality and subjective audio-video delay tolerance. In *MOSAIC Handbook*, pp. 183–187.

Robson, J. G. (1966). Spatial and temporal contrast-sensitivity functions of the visual system. *Journal of the Optical Society of America* **56**:1141–1142.

Rogowitz, B. E. (1983). The human visual system: A guide for the display technologist. In *Proceedings of the SID*, **24**:235–252.

Rohaly, A. M., Ahumada, A. J. Jr, Watson, A. B. (1997). Object discrimination in natural background predicted by discrimination performance and models. *Vision Research* **37**(23):3225–3235.

Rohaly, A. M. *et al.* (2000). Video Quality Experts Group: Current results and future directions. In *Proc. SPIE Visual Communications and Image Processing*, vol. 4067, pp. 742–753, Perth, Australia.

Ross, J., Speed, H. D. (1991). Contrast adaptation and contrast masking in human vision. *Proceedings of the Royal Society of London B* **246**:61–70.

Roufs, J. A. J. (1989). Brightness contrast and sharpness, interactive factors in perceptual image quality. In *Proc. SPIE Human Vision, Visual Processing and Digital Display*, vol. 1077, pp. 209–216, Los Angeles, CA.

Roufs, J. A. J. (1992). Perceptual image quality: Concept and measurement. *Philips Journal of Research* **47**(1):35–62.

Rovamo, J., Kukkonen, H., Mustonen, J. (1998). Foveal optical modulation transfer function of the human eye at various pupil sizes. *Journal of the Optical Society of America A* **15**(9):2504–2513.

Salembier, P., Marqués, F. (1999). Region-based representations of image and video: Segmentation tools for multimedia services. *IEEE Transactions on Circuits and Systems for Video Technology* **9**(8):1147–1169.

Savakis, A. E., Etz, S. P., Loui, A. C. (2000). Evaluation of image appeal in consumer photography. In *Proc. SPIE Human Vision and Electronic Imaging*, vol. 3959, pp. 111–120, San Jose, CA.

Sayood, K. (2000). *Introduction to Data Compression*, 2nd edn, Morgan Kaufmann.

Schade, O. H. (1956). Optical and photoelectric analog of the eye. *Journal of the Optical Society of America* **46**(9):721–739.

Sekuler, R., Blake, R. (1990). *Perception*, 2nd edn, McGraw-Hill.

Seyler, A. J., Budrikis, Z. L. (1959). Measurements of temporal adaptation to spatial detail vision. *Nature* **184**:1215–1217.

Seyler, A. J., Budrikis, Z. L. (1965). Detail perception after scene changes in television image presentations. *IEEE Transactions on Information Theory* **11**(1):31–43.

Simoncelli, E. P., Freeman, W. T., Adelson, E. H., Heeger, D. J. (1992). Shiftable multi-scale transforms. *IEEE Transactions on Information Theory* **38**(2):587–607.

Snowden, R. J., Hammett, S. T. (1996). Spatial frequency adaptation: Threshold elevation and perceived contrast. *Vision Research* **36**(12):1797–1809.

Stein, E. M., Weiss, G. (1971). *Introduction to Fourier Analysis on Euclidean Spaces*, Princeton University Press.

Steinmetz, R. (1996). Human perception of jitter and media synchronization. *IEEE Journal on Selected Areas in Communications* **14**(1):61–72.

Stelmach, L. B., Tam, W. J. (1994). Processing image sequences based on eye movements. In *Proc. SPIE Human Vision, Visual Processing and Digital Display*, vol. 2179, pp. 90–98, San Jose, CA.

Stelmach, L. B., Tam, W. J., Hearty, P. J. (1991). Static and dynamic spatial resolution in image coding: An investigation of eye movements. In *Proc. SPIE Human Vision, Visual Processing and Digital Display*, vol. 1453, pp. 147–152, San Jose, CA.

Stockman, A., Sharpe, L. T. (2000). Spectral sensitivities of the middle- and long-wavelength sensitive cones derived from measurements in observers of known genotype. *Vision Research* **40**(13):1711–1737.

Stockman, A., MacLeod, D. I. A., Johnson, N. E. (1993). Spectral sensitivities of the human cones. *Journal of the Optical Society of America A* **10**(12):2491–2521.

Stockman, A., Sharpe, L. T., Fach, C. (1999). The spectral sensitivity of the human short-wavelength sensitive cones derived from thresholds and color matches. *Vision Research* **39**(17):2901–2927.

Stromeyer III, C. F., Klein, S. (1975). Evidence against narrow-band spatial frequency channels in human vision: The detectability of frequency modulated gratings. *Vision Research* **15**:899–910.

Süsstrunk, S., Winkler, S. (2004). Color image quality on the Internet. In *Proc. SPIE Internet Imaging*, vol. 5304, pp. 118–131, San Jose, CA (invited paper).

Svaetichin, G. (1956). Spectral response curves from single cones. *Acta Physiologica Scandinavica* **134**:17–46.

Switkes, E. Bradley, A. De Valois, K. K., (1988). Contrast dependence and mechanisms of masking interactions among chromatic and luminance gratings. *Journal of the Optical Society of America A* **5**(7):1149–1162.

Symes, P. (2003). *Digital Video Compression*, McGraw-Hill.

Tam, W. J. *et al.* (1995). Visual masking at video scene cuts. In *Proc. SPIE Human Vision, Visual Processing and Digital Display*, vol. 2411, pp. 111–119, San Jose, CA.

Tan, K. T., Ghanbari, M., Pearson, D. E. (1998). An objective measurement tool for MPEG video quality. *Signal Processing* **70**(3):279–294.

Teo, P. C., Heeger, D. J. (1994a). Perceptual image distortion. In *Proc. SPIE Human Vision, Visual Processing and Digital Display*, vol. 2179, pp. 127–141, San Jose, CA.

Teo, P. C., Heeger, D. J. (1994b). Perceptual image distortion. In *Proc. International Conference on Image Processing*, vol. 2, pp. 982–986, Austin, TX.

Thomas, G. (1998). A comparison of motion-compensated interlace-to-progressive conversion methods. *Signal Processing: Image Communication* **12**(3):209–229.

Tong, X., Heeger, D., van den Branden Lambrecht, C. J. (1999). Video quality evaluation using ST-CIELAB. In *Proc. SPIE Human Vision and Electronic Imaging*, vol. 3644, pp. 185–196, San Jose, CA.

Tudor, P. N. (1995). MPEG-2 video compression. *Electronics & Communication Engineering Journal* **7**(6):257–264.

van den Branden Lambrecht, C. J. (1996a). Color moving pictures quality metric. In *Proc. International Conference on Image Processing*, vol. 1, pp. 885–888, Lausanne, Switzerland.

van den Branden Lambrecht, C. J. (1996b). *Perceptual Models and Architectures for Video Coding Applications*. PhD thesis, École Polytechnique Fédérale de Lausanne, Switzerland.

van den Branden Lambrecht, C. J., Farrell, J. E. (1996). Perceptual quality metric for digitally coded color images. In *Proc. European Signal Processing Conference*, pp. 1175–1178, Trieste, Italy.

van den Branden Lambrecht, C. J., Verscheure, O. (1996). Perceptual quality measure using a spatio-temporal model of the human visual system. In *Proc. SPIE Digital Video Compression: Algorithms and Technologies*, vol. 2668, pp. 450–461, San Jose, CA.

van den Branden Lambrecht, C. J., Costantini, D. M., Sicuranza, G. L., Kunt, M. (1999). Quality assessment of motion rendition in video coding. *IEEE Transactions on Circuits and Systems for Video Technology* **9**(5):766–782.

van Hateren, J. H., van der Schaaf, A. (1998). Independent component filters of natural images compared with simple cells in primary visual cortex. *Proceedings of the Royal Society of London B* **265**:1–8.

Vandergheynst, P., Gerek, Ö. N. (1999). Nonlinear pyramidal image decomposition based on local contrast parameters. In *Proc. Nonlinear Signal and Image Processing Workshop*, vol. 2, pp. 770–773, Antalya, Turkey.

Vandergheynst, P., Kutter, M., Winkler, S. (2000). Wavelet-based contrast computation and its application to watermarking. In *Proc. SPIE Wavelet Applications in Signal and Image Processing*, vol. 4119, pp. 82–92, San Diego, CA (invited paper).

Vimal, R. L. P. (1997). Orientation tuning of the spatial-frequency mechanisms of the red-green channel. *Journal of the Optical Society of America A* **14**(10):2622–2632.

VQEG (2000). Final report from the Video Quality Experts Group on the validation of objective models of video quality assessment. Available at http://www.vqeg.org/

VQEG, (2003). Final report from the Video Quality Experts Group on the validation of objective models of video quality assessment – Phase II. Available at http://www.vqeg.org/

Wandell, B. A. (1995). *Foundations of Vision*, Sinauer Associates.

Wang, Y., Zhu, Q.-F. (1998). Error control and concealment for video communications: A review. *Proceedings of the IEEE* **86**(5):974–997.

Wang, Z., Sheikh, H. R., Bovik, A. C. (2002). No-reference perceptual quality assessment of JPEG compressed images. In *Proc. International Conference on Image Processing*, vol. 1, pp. 477–480, Rochester, NY.

Watson, A. B. (1986). Temporal sensitivity. In Boff, K. R., Kaufman, L., Thomas, J. P. (eds), *Handbook of Perception and Human Performance*, vol. 1, chap. 6, John Wiley.

Watson, A. B. (1987a). The cortex transform: Rapid computation of simulated neural images. *Computer Vision, Graphics, and Image Processing* **39**(3):311–327.

Watson, A. B. (1987b). Efficiency of a model human image code. *Journal of the Optical Society of America A* **4**(12):2401–2417.

Watson, A. B. (1990). Perceptual-components architecture for digital video. *Journal of the Optical Society of America A* **7**(10):1943–1954.

Watson, A. B. (1995). Image data compression having minimum perceptual error. US Patent 5,426,512.

Watson, A. B. (1997). Image data compression having minimum perceptual error. US Patent 5,629,780.

Watson, A. B. (1998). Toward a perceptual video quality metric. In *Proc. SPIE Human Vision and Electronic Imaging*, vol. 3299, pp. 139–147, San Jose, CA.

Watson, A. B., Ahumada, A. J. Jr. (1989). A hexagonal orthogonal-oriented pyramid as a model of image representation in visual cortex. *IEEE Transactions on Biomedical Engineering* **36**(1):97–106.

Watson, A. B., Pelli, D. G. (1983). QUEST: A Bayesian adaptive psychometric method. *Perception & Psychophysics* **33**(2):113–120.

Watson, A. B., Solomon, J. A. (1997). Model of visual contrast gain control and pattern masking. *Journal of the Optical Society of America A* **14**(9):2379–2391.

Watson, A. B., Borthwick, R. Taylor, M. (1997). Image quality and entropy masking. In *Proc. SPIE Human Vision and Electronic Imaging*, vol. 3016, pp. 2–12, San Jose, CA.

Watson, A. B., Hu, J., McGowan III, J. F., Mulligan, J. B. (1999). Design and performance of a digital video quality metric. In *Proc. SPIE Human Vision and Electronic Imaging*, vol. 3644, pp. 168–174, San Jose, CA.

Watson, A. B., Hu, J., McGowan III, J. F. (2001). Digital video quality metric based on human vision. *Journal of Electronic Imaging* **10**(1), pp. 20–29.

Webster, M. A., Miyahara, E. (1997). Contrast adaptation and the spatial structure of natural images. *Journal of the Optical Society of America A* **14**(9):2355–2366.

Webster, M. A., Mollon, J. D. (1997). Adaptation and the color statistics of natural images. *Vision Research* **37**(23):3283–3298.

Webster, M. A., De Valois, K. K., Switkes, E. (1990). Orientation and spatial-frequency discrimination for luminance and chromatic gratings. *Journal of the Optical Society of America A* **7**(6):1034–1049.

Weibull, W. (1951). A statistical distribution function of wide applicability. *Journal of Applied Mechanics* **18**:292–297.

Westen, S. J. P., Lagendijk, R. L., Biemond, J. (1997). Spatio-temporal model of human vision for digital video compression. In *Proc. SPIE Human Vision and Electronic Imaging*, vol. 3016, pp. 260–268, San Jose, CA.

Westerink, J. H. D. M., Roufs, J. A. J. (1989). Subjective image quality as a function of viewing distance, resolution, and picture size. *SMPTE Journal* **98**(2):113–119.

Westheimer, G. (1986). The eye as an optical instrument. In Boff, K. R., Kaufman, L., Thomas J. P. (eds), *Handbook of Perception and Human Performance*, vol. 1, chap. 4, John Wiley.

Williams, D. R., Brainard, D. H., McMahon, M. J., Navarro, R. (1994). Double-pass and interferometric measures of the optical quality of the eye. *Journal of the Optical Society of America A* **11**(12):3123–3135.

Wilson, H. R., Humanski, R. (1993). Spatial frequency adaptation and contrast gain control. *Vision Research* **33**(8):1133–1149.

Winkler, S. (1998). A perceptual distortion metric for digital color images. In *Proc. International Conference on Image Processing*, vol. 3, pp. 399–403, Chicago, IL.

Winkler, S. (1999a). Issues in vision modeling for perceptual video quality assessment. *Signal Processing* **78**(2):231–252.

Winkler, S. (1999b). A perceptual distortion metric for digital color video. In *Proc. SPIE Human Vision and Electronic Imaging*, vol. 3644, pp. 175–184, San Jose, CA.

Winkler, S. (2000). Quality metric design: A closer look. In *Proc. SPIE Human Vision and Electronic Imaging*, vol. 3959, pp. 37–44, San Jose, CA.

Winkler, S. (2001). Visual fidelity and perceived quality: Towards comprehensive metrics. In *Proc. SPIE Human Vision and Electronic Imaging*, vol. 4299, pp. 114–125, San Jose, CA.

Winkler, S., Campos, R. (2003). Video quality evaluation for Internet streaming applications. In *Proc. SPIE Human Vision and Electronic Imaging*, vol. 5007, pp. 104–115, Santa Clara, CA.

Winkler, S., Dufaux, F. (2003). Video quality evaluation for mobile applications. In *Proc. SPIE Visual Communications and Image Processing*, vol. 5150, pp. 593–603, Lugano, Switzerland.

Winkler, S., Faller, C. (2005). Audiovisual quality evaluation of low-bitrate video. In *Proc. SPIE Human Vision and Electronic Imaging*, vol. 5666, San Jose, CA.

Winkler, S., Sharma, A., McNally, D. (2001). Perceptual video quality and blockiness metrics for multimedia streaming applications. In *Proc. International Symposium on Wireless Personal Multimedia Communications*, pp. 553–556, Aalborg, Denmark (invited paper).

Winkler, S., Süsstrunk, S. (2004). Visibility of noise in natural images. In *Proc. SPIE Human Vision and Electronic Imaging*, vol. 5292, pp. 121–129, San Jose, CA.

Winkler, S., Vandergheynst, P. (1999). Computing isotropic local contrast from oriented pyramid decompositions. In *Proc. International Conference on Image Processing*, vol. 4, pp. 420–424, Kyoto, Japan.

Wolf, S., Pinson, M. H. (1999). Spatial-temporal distortion metrics for in-service quality monitoring of any digital video system. In *Proc. SPIE Multimedia Systems and Applications*, vol. 3845, pp. 266–277, Boston, MA.

Wyszecki, G., Stiles, W. S. (1982). *Color Science: Concepts and Methods, Quantitative Data and Formulae*, 2nd edn, John Wiley.

Yang, J., Makous, W. (1994). Spatiotemporal separability in contrast sensitivity. *Vision Research* **34**(19):2569–2576.

Yang, J., Makous, W. (1997). Implicit masking constrained by spatial inhomogeneities. *Vision Research* **37**(14):1917–1927.

Yang, J., Lu, W., Waibel, A. (1998). Skin-color modeling and adaptation. In *Proc. Asian Conference on Computer Vision*, vol. 2, pp. 687–694, Hong Kong.

Yendrikhovskij, S. N., Blommaert, F. J. J., de Ridder, H. (1998). Perceptually optimal color reproduction. In *Proc. SPIE Human Vision and Electronic Imaging*, vol. 3299, pp. 274–281, San Jose, CA.

Young, R. A. (1991). Oh say, can you see? The physiology of vision. In *Proc. SPIE Human Vision, Visual Processing and Digital Display*, vol. 1453, pp. 92–123, San Jose, CA.

Yu, Z., Wu, H. R., Chen, T. (2000). A perceptual measure of ringing artifact for hybrid MC/DPCM/DCT coded video. In *Proc. IASTED International Conference on Signal and Image Processing*, pp. 94–99, Las Vegas, NV.

Yu, Z., Wu, H. R., Winkler, S., Chen, T. (2002). Vision model based impairment metric to evaluate blocking artifacts in digital video. *Proceedings of the IEEE* **90**(1):154–169.

Yuen, M., Wu, H. R. (1998). A survey of hybrid MC/DPCM/DCT video coding distortions. *Signal Processing* **70**(3):247–278.

Zhang, X., Wandell, B. A. (1996). A spatial extension of CIELAB to predict the discriminability of colored patterns. In *SID Symposium Digest*, vol. 27, pp. 731–735.

Ziliani, F. (2000). *Spatio-Temporal Image Segmentation: A New Rule-Based Approach*. PhD thesis, École Polytechnique Fédérale de Lausanne, Switzerland.

Index

Digital Video Quality - Vision Models and Metrics   Stefan Winkler
© 2005 John Wiley & Sons, Ltd   ISBN: 0-470-02404-6